SEA OTTERS

Studies in Pacific Worlds

SERIES EDITORS
Rainer F. Buschmann
Katrina Gulliver

EDITORIAL BOARD
Gregory Cushman
John Gascoigne
Andrea Geiger
Noriko Kawamura
Judith Schachter

"[*Sea Otters*] provides readers with a fascinating overview of the life, times, and history of the smallest marine mammal in the North Pacific Ocean. It is at once natural history, commercial history, imperial and nation defining history, species extinction history, conservation history, and tourism/entertainment history."
—Robin Inglis, *Ormsby Review*

"Expertly integrating history and biology, this is the one book that tells the full, tragic story of the sea otter from its near extinction to its elevation to icon of cuteness. The sea otter, as Ravalli masterfully relates, has long been at the center of politics, conservation, and tourism in the North Pacific. Before you visit the sea otters at a Pacific aquarium, read this book to understand the fascinating history of how these creatures got there, and how they very nearly did not make it."
—Ryan Tucker Jones, associate professor of history at the University of Oregon

"Well researched and succinctly told, this is the story of the late eighteenth-century sea otter trade that decimated a unique marine species and revolutionized the Pacific Rim by introducing coastal communities to a global capitalist system."
—Jim Hardee, editor of *The Rocky Mountain Fur Trade Journal*

"Here is the story, richly told, of how these vulnerable mammals—the ermine of Asian markets—were pursued for their lustrous skins and hunted to near extinction. The quest eventually generated a rivalry between seafaring nations and indigenous peoples along islands and coasts from China to Mexico."
—Barry Gough, professor emeritus of history at Wilfrid Laurier University and author of *Pax Britannica: Ruling the Waves and Keeping the Peace before Armageddon*

Sea Otters

A HISTORY

Richard Ravalli

UNIVERSITY OF NEBRASKA PRESS • LINCOLN

© 2018 by the Board of Regents of
the University of Nebraska

Acknowledgments for the use of copyrighted
material appear on page xii, which constitutes
an extension of the copyright page.

All rights reserved

Library of Congress Cataloging-in-Publication Data
Names: Ravalli, Richard, author.
Title: Sea otters: a history / Richard Ravalli.
Description: Lincoln: University of Nebraska
Press, [2018] | Series: Studies in Pacific worlds |
Includes bibliographical references and index.
Identifiers: LCCN 2018003978
ISBN 9780803284401 (cloth: alk. paper)
ISBN 9781496225009 (paperback)
ISBN 9781496212184 (epub)
ISBN 9781496212191 (mobi)
ISBN 9781496212207 (pdf)
Subjects: LCSH: Sea otter—Pacific Area—History. |
Sea otter populations—Pacific Area—History. | Sea
otter—Conservation—Pacific Area—History.
Classification: LCC QL737.C25 R395 2018 |
DDC 333.95/97695—dc23 LC record available
at https://lccn.loc.gov/2018003978

Set in New Baskerville ITC by Mikala R. Kolander.
Designed by N. Putens.

for my wife, Lisa, daughters, Rachel and Sarah, Mom, Dad,

family, and students, who all kept me afloat

CONTENTS

List of Illustrations   *ix*

Acknowledgments   *xi*

Introduction   *xiii*

1. Rakkoshima, the Sea Otter Islands   *1*
2. *Promyshlenniki* and Padres   *21*
3. Boston Men   *47*
4. Near Extinction and Reemergence   *77*
5. Nukes, Aquaria, and Cuteness   *103*

    Conclusion   *125*

    Appendix: List of Vessels Engaged in the California Sea Otter Trade, 1786–1847   *127*

    Notes   *141*

    Bibliography   *161*

    Index   *179*

LIST OF ILLUSTRATIONS

1. Steller's sea otter from *De bestiis marinis* xx
2. Woodblock of Ainu hunting sea otters 5
3. Woodblock of Aleuts hunting sea otters 18
4. Painting of a sea otter from George Shaw's *Musei Leveriani Explication* 29
5. Map from Pedro Calderón y Henríquez manuscript 31
6. Graph of Russian sea otter pelt exports from the Russian North Pacific, 1742–1797 42
7. Graph of California sea otter pelt exports, 1786–1847 42
8. Painting of *"Enhydra marina"* by John Woodhouse Audubon 76
9. Painting of the schooner *Cygnet* 80
10. Dead sea otter near Oyhut, Washington 85
11. California wardens with pup 102
12. Robert "Sea Otter" Jones with sea otter 106
13. Sea otter at Monterey Bay Aquarium 116

ACKNOWLEDGMENTS

I have many people to thank for this book. My professors at California State University, Stanislaus, where this research began, encouraged and mentored me early on, especially Nancy Taniguchi, Richard Weikart, and Bob Santos. At the University of California, Merced, Gregg Herken offered a steady, guiding hand and gave me the freedom to venture into unfamiliar waters. (Fellow founding UCM graduate students Bradford Johnston, Trevor Albertson, Ray Winter, and Joel Beutel also deserve recognition for their friendship.) David Igler at the University of California, Irvine, graciously read my chapters from afar and invited me to conferences, lonely as I was in the Central Valley. Nothing could have been accomplished without Gregg and David, and I owe them both a great deal of gratitude. I also thank Ryan Tucker Jones for his expertise and support, as well as my colleagues Rex Gurney and Michael McGrann.

The staffs at the Waseda University Library, the Huntington Library, the Bancroft Library, the Holt-Atherton Special Collections at the University of the Pacific, the California State Library, and the Monterey Public Library (especially archivist Dennis Copeland) offered kind assistance. Thanks must also be given to librarians Kacey Bullock and Peter Gabbani at Sierra College and Andrew Tweet at Folsom Public Library for their help in tracking down sources. A few students provided research assistance for this book. My sincerest thanks to Hannah Jones, Kirsten Livingston, and Jason Peters for their help and for motivating me to finish this project.

I cannot praise Bridget Barry and Emily Wendell at the University of Nebraska Press enough. Individuals whom I only know through correspondence provided gracious support along the way. Marla Daily at the Santa Cruz Island Foundation was helpful at the later stages of this project. In Alaska thanks go to J. Pennelope Goforth, Peter Williams, and Diane Purvis. In Canada I thank Maureen Weddell at the Haida Gwaii Museum, Cynthia Holmes for talking with me about her video of Nyac and Milo, and Barry Gough for his kind words of encouragement. I hope to travel the Pacific and meet some of you soon.

Finally, I thank my parents, Richard and Anna, for taking me, Frankie, Julie, and Robi on all those trips to the California coast. Lisa, Rachel, and Sarah suffered their own car rides to the coast, as well as the hours that I locked myself away. This book is for family.

I wish to thank the following publications for their support of my work.

Chapter 1 was originally published as "The Sea Otter Islands: Geopolitics and Environment in the East Asian Fur Trade," *Asia Pacific: Perspectives* 9, no. 2 (June 2010): 27–34.

Chapter 4 was published as "The Near Extinction and Reemergence of the Pacific Sea Otter, 1850–1938," *Pacific Northwest Quarterly* 100, no. 4 (Fall 2009): 181–91.

Chapter 5 was originally published as "Sea Otter Aesthetics in Popular Culture," in *Animals in Human Society: Amazing Creatures Who Share Our Planet*, edited by Daniel Moorehead, 93–104 (Lanham MD: University Press of America, 2016).

An earlier version of the appendix was published with Kirsten Livingston and Hannah Zimmerman, "A Revised List of Vessels Engaged in the California Sea Otter Trade, 1786–1847," *International Journal of Maritime History* 24, no. 2 (December 2012): 225–38.

INTRODUCTION

The sea otter is one of the most important nonhuman species in Pacific history. Pursuit of the animals in the early modern era was responsible for the intensification of foreign activity in the globe's largest ocean. Around three centuries ago, sea otter fur helped put the Pacific on the map. As rival empires vied for control of the maritime fur trade, they reshaped geopolitical boundaries in the region. As hunters killed sea otters, they brought them to near extinction throughout the otters' coastal ranges, and in some places humans eliminated them. This is a story of global trade, imperial conflict, environmental tragedy, and the survival and resilience of an iconic marine mammal.

General outlines of the sea otter's past have been provided in a variety of sources on the fur trade, the North American West, and natural history. What has been missing until now is a book focused on sea otters and people since early modern times. This book fills that gap. Previous academic studies have emphasized the maritime fur trade in particular areas such as the Pacific Northwest and California or specific national participants.[1] *Sea Otters* offers a synthesis of these approaches and contributes to them in a number of ways. It seeks to expand the history of the trade both geographically and temporally. Relatively little on sea otter hunting and trading in Asia and the western Pacific has been written by historians, yet it was otter pelts from Hokkaido and the Kuril Islands, not Vancouver Island in British Columbia, that first reached Japanese and Chinese consumers as luxury goods. These broader global dynamics are incorporated here. Additionally, prior authors have been largely silent on sea otter hunting, trading, and conservation following the waning of the trade in the eastern Pacific at the middle of the nineteenth century. Biologists and conservationists have helped to tell some of these stories and have contributed to historical research on sea otters into the twentieth and twenty-first centuries. *Sea Otters* merges these different perspectives to illuminate new places and periods in the animal's history while drawing upon insight from multiple disciplines.

This book is inspired by recent analyses of the Pacific World.[2] Within the last decade and a half, historians have produced an increased number of works that frame events and developments within expansive oceanic contexts and emphasize how the Pacific is an important, yet often neglected, region. Thinking about sea otters in this way makes sense in part because they are only found in nearshore habitat along the coastlines of the North Pacific. Also, an expanded geographic scope that encompasses the entire range of the species allows for comparisons and contrasts of hunting and conservation across seascapes. It encourages us to analyze links between specific locations in the Pacific where the creatures lived and died. When U.S. hunters based in California pursued sea otters in the Kuril Islands in the late nineteenth century, their actions helped ensure that the remaining California population would survive—although barely—into the twentieth century. Other connections examined in this volume demonstrate that the places where otters and humans interacted were not isolated from each other. They were joined by the same commercial, geopolitical, and environmental dynamics that had disparate effects on the animals throughout their marine home. Sea otters not only helped to bring about this increasingly interconnected Pacific World in the eighteenth century but are shaped by the forces that have defined it since then.

Sea otters have been carried by the ocean's political currents. Various national and international conflicts and resolutions in which they have been entangled over the last few hundred years are emphasized in the following pages. The fur trade was closely connected with Russian imperial expansions in the North Pacific in the eighteenth century, activities that eventually encouraged Spain to consolidate and fortify coastal territory north of Mexico. British merchants followed in the wake of

James Cook's third expedition and challenged Spanish title to Nootka Sound on the western coast of Vancouver Island. These colonial and diplomatic showdowns are well known to scholars, as is the realization that they were often resolved in distant capitals without regard for the sovereignty of the Pacific's indigenous people. Yet the political value of the sea otter runs deep and is difficult to separate from its economic worth. Historian David Igler argues that the imperial contexts of the eighteenth- and nineteenth-century Pacific have been exaggerated in historiography that ignores local complexities and fails to appreciate that American sailors and merchants (who were the majority of non-Natives engaged in the region's maritime enterprises by the early 1800s) were engaged in entrepreneurial activity, not flag waving.[3] Still, U.S. expansion to the far west coast in the 1840s did not emerge in a vacuum, and prominent American fur traders at that time, men like John Jacob Astor and William Sturgis, recognized the national importance of their commerce even as their ambitions were mitigated by the designs of international rivals. To be sure, most individuals were too busy making money in the Pacific to worry about imperial claims. Nevertheless, otter pelts were a marine resource imbued often enough by terrestrial outsiders and maritime sojourners with both commercial and territorial significance. *Sea Otters* tries to keep this significance in focus.

Land claims involving skins meant a lot to the marine mammal bodies to which both were attached. Hence the environmental dimensions of sea otter hunting and conservation, which historians have begun to pay more attention to in recent years, is another focus of this book.[4] Intensified human predation beginning in the eighteenth century dramatically reduced the number of animals in the Pacific, but this decline took place at different rates and at different places and times. Moreover, sea otter conservation presents a more complicated story than histories of the species usually suggest. For example, the twentieth century is not the first time that otters recovered from localized depletions caused by hunting. As we will see, Russian authorities played a role in helping otters rebound in the nineteenth century. Additionally, many may not realize that killer whales were great killers of sea otters at the turn of the millennium and threatened their recovery in Alaska or that the popular image of the creatures as cute and cuddly is not always appreciated by conservationists. These complexities are examined in the following chapters. *Sea Otters* sheds new light on marine mammal history and encourages dialogue between and partnership among scientists, humanists, and advocates.

To this end, each of the following chapters is indebted to information provided by relevant academic approaches. Susan Nance has called upon historians to "employ a radically interdisciplinary supporting literature" in the area of animal history, and her advice encourages this study.[5] Chapter 1 establishes a broader geographic scope for the maritime fur trade. Emphasizing sea otter hunting and exchange networks in the western Pacific opens up new trails and allows for a synthesis of insights from historians in various subfields. The chapter likewise consults biologists, archaeologists, and others to illuminate the environmental contexts of the East Asian otter trade. While all sections of *Sea Otters* are limited to English-language sources, researchers working in Russian, Japanese, and Chinese archives can build upon the groundwork established in the following examination of the "Sea Otter Islands." Chapter 2 looks at sea otter history after marine mammal hunting and trading moved from contested Russian and Japanese waters to the eastern Pacific at the middle of the eighteenth century. It informs the work of historians and marine mammal biologists by highlighting revised data on pelt cargoes. The damage done to sea otter populations by Russian ventures in the eighteenth and nineteenth centuries was real, and more exact comparative knowledge of these changes is needed to better understand the effects of maritime exploitation.

Similarly, chapter 3 references archival data relating to California sea otter exports to better estimate past population declines. It builds in part upon a study by historical ecologists and anthropologists who used previously published data to illustrate pelt shipments from Alta and Baja California prior to the gold rush.[6] Along with providing updated information, the chapter places otter decline through the first half of the nineteenth century in geographic perspective. Thus hunting along the coasts of modern-day Washington State and Oregon is considered in relation to California. In the late 1800s and into the twentieth century, sea otters across the Pacific were overhunted, and in a number of locations they were extirpated. Chapter 4 details these underexplored transoceanic developments by synthesizing historical and scientific literature. In particular, sources in marine mammal science help to tell rich stories of *Enhydra*'s near extinction and early twentieth-century conservation and highlight opportunities for collaboration. If Russians were both friends and enemies of the species before 1900, they thereafter built dramatically on the former by engaging in the first captivity and translocation studies during the 1930s, one of a number of topics that needs more attention from historians. Lastly, this book explores sea otter history and conservation issues since the middle of the twentieth century. Chapter 5 develops an earlier analysis of otter aesthetics and culture that was included in an interdisciplinary animal studies volume. In addition to cultural concerns, it identifies political and environmental events that have substantially reshaped the contemporary world of sea otters.[7]

A Prehistory

Around five million years ago, the animal *Enhydritherium* migrated from Eurasia to North America, spreading from the Old World to the New—as one species designation, *terraenovae*, suggests—and to the western coast of North America. The recent discovery of a fossil tooth in central Mexico suggests that *Enhydritherium* may have crossed westward over the continent utilizing smaller bodies of water and a less circuitous route than via Canada or Panama. Unlike its relative *Enhydra lutris*, the sea otter, it could live in both freshwater and marine environments but was similar in body size. *Enhydritherium* is thus an extinct Pacific cousin of the sea otter, the latter appearing in Pleistocene fossil records from one to three million years ago and only in the North Pacific. The precise evolutionary relationship between them and how *Enhydra* dispersed throughout the ocean is still debated.[8] As the new creatures colonized the Pacific, sea otters were blocked by ice in far northern reaches and by tropical conditions farther south, their biological adaptations limiting them to a nearshore latitudinal band stretching from Hokkaido and Sakhalain Islands in the west to Baja California in the east. They became shellfish eaters possibly due to competition with pinnipeds and other fish-eating species.[9]

Otters interacted with a rich and dynamic natural landscape in the North Pacific for millennia before contact with humans. The region is home to one of the most diverse assortments of marine life on the globe, including most varieties of seals and sea lions, as well as the now-extinct sea cow. Currents and freshwater outlets contribute to an oceanographic process known as "upwelling," or the rising of cold, nutrient-rich water from deep in the Pacific within which plankton abound. This process supports abundant fish, bird, and mammal populations, which appear to thrive during periods of colder sea and air temperatures.[10] Sea otters adapted to consuming macroinvertebrates in part through the development of tool use, and this ability has had dramatic effects on nearshore ecologies, something that scientists began to understand more clearly by the late twentieth century. Predation of sea urchins, and urchin predation of kelps, mean that hungry otters support vibrant kelp populations by limiting the number of herbivores in particular locales. Kelp forests in turn benefit a variety of other species, linking *Enhydra* in a trophic cascade across its range. As biologist James Estes notes, these relationships also worked to suppress an "evolutionary arms race" between kelps and herbivores and may help explain why Pacific kelp is especially vulnerable to sea urchin grazing in the absence of sea otters.[11] A curious form of environmental détente across time and seascape can be added to the record of these fascinating aquatic mammals.

Unlike whales and pinnipeds, sea otters lack a layer of blubber to protect them from the cold ocean and rely instead on the densest fur of any species on earth. Their coats contain up to one million hairs per square inch, with an undercoat and longer guard hairs that effectively lock in air to insulate their bodies. This soft, rich pelage continues to inspire people to kill them, yet the most immediate threat is heat loss; thus sea otters spend about 10 percent of each day grooming to prevent contamination of their fur. A high metabolic rate also keeps their bodies warm and makes them prodigious eaters. Sea otters spend up to half of their day foraging and consume as much as a quarter of their body weight.[12] Females give birth to one pup at a time, and in rare instances when more than one pup is born, only a single pup survives infancy. The otter's attractive coat and its slow reproductive rate have made the species vulnerable to hunting pressures, since the result of those pressures is relatively slow population recovery. Other than humans, the only habitual predators of sea otters are killer whales, great white sharks, bald eagles, and arctic foxes.[13] Together they pursued some 150,000 to 300,000 individuals prior to the modern fur trade.[14]

When not diving for food or traveling, sea otters congregate in large groups, or "rafts." Males and females are most often separated in different rafts, with females staying in the same area for days or years and males tending to be more mobile.[15] While some historical evidence suggests that they more frequently hauled out on rocks and beaches prior to the intensive hunting of the eighteenth and nineteenth centuries, sea otters today spend most of their time in the water.[16] Preferring to float on their backs, otters are known to wrap themselves in kelp to keep from drifting and will on rare occasions "hold hands." These and other mannerisms are often perceived as humanlike by observers. In the first European description of the sea otter, published in 1751, German naturalist Georg Wilhelm Steller wrote, "They play together, and, like human beings, embrace with their arms and kiss each other."[17] Indigenous people also found it difficult to resist thinking about otters in such romantic terms. The anthropological evidence is further indication of a link between hunting sea otters and romanticizing them. Those humans who did not chase otters into the surf and encourage them to showcase their aquatic charisma appear to have been less interested in considering otters spirit beings or kin.

The littoral environment is where sea otters eat, mate, rear young, shelter, and play. For Igler, thinking about the edges of the Pacific is useful for understanding spaces where "a great deal of history has taken place." Oceanic and terrestrial forces shaped coastal zones of Earth's largest ocean and often had dramatic effects on local flora and fauna

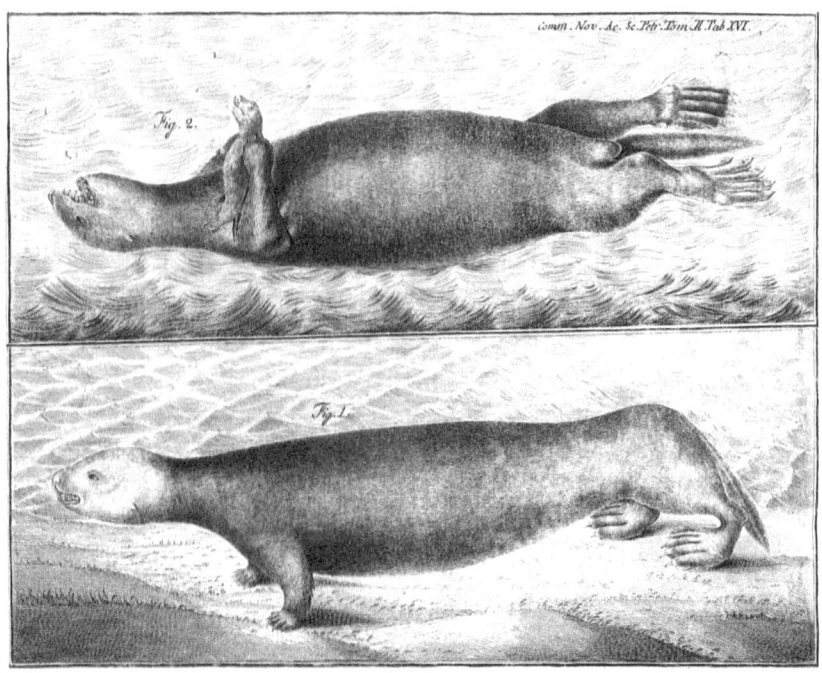

FIG. 1. This drawing accompanied the first European description of the sea otter in Georg Steller's *De bestiis marinis* in 1751. The unnatural look of the animal here may have influenced some later depictions. Wikimedia Commons.

that were "hardly Pacific."[18] Thus it was in these amphibious geographies where eagles preyed on sea otter pups, sharks snatched their parents from the surface, and the first person ended the life of a sea otter. Perhaps this occurred tens of thousands of years ago around the time human migrants crossed to the Japanese islands from northeastern Asia and Siberia. After 10,000 B.C. Japan's ancient Jomon culture left evidence of sea otter hunting at several archaeological sites on Hokkaido as marine mammal exploitation in the Sea of Okhotsk increased over time.[19] As the first Amerindians traversed Bering Strait land bridges and followed Pleistocene animals down an ice-free corridor near the Rocky Mountains, hunters utilized marine mammal and other coastal resources to help people the Americas. Some of the earliest evidence for sea otter hunting in the eastern Pacific comes from Haida Gwaii off the British Columbia coast at intertidal sites dated to around 8000 B.C. Ancestors of the Chumash on California's Channel Islands used watercraft to pursue *Enhydra lutris* and left its bones behind by at least a millennium later.[20] The littoral became an increasingly dangerous space for sea otters in prehistory, and—foreshadowing what would unfold during the maritime fur trade—the species played a role in attracting the attention of outsiders to the Pacific World.

The question of indigenous overexploitation of sea otters has received attention in recent years from archaeologists and paleoecologists. Pacific maritime cultures reshaped natural environments in order to sustain life in often inhospitable locales. For example, Natives of the volcanic Aleutian Islands, with little in the way of agricultural or terrestrial resources, turned to the ocean for food and clothing and developed a number of methods for harvesting marine mammals.[21] Some scholars contend that Aleut hunting of sea otters (and subsequent exploitation of sea urchins in the absence of otters) is linked to the existence of a mosaic of kelp forests and barrens in the region over thousands of years and that while localized extinctions did take place, the ecosystem was overall in a state of equilibrium. Depletion of otter populations by the Chumash after 5000 B.C. may be related to an increase in abalone harvesting in the Channel Islands, both of which resulted in alterations to kelp forests and a rise in sea urchin numbers.[22] However, most other California coastal tribes, with the exception of the Humalgueno Indians of Baja, did not engage in extensive hunting of marine mammals.[23]

Additional studies have examined historical ecologies in the eastern Pacific to better inform current conservation efforts and to engage longstanding debates about early Native Americans and the environment. Shell middens (ancient garbage piles) at Sanak Island in the Aleutians confirm that inhabitants there processed sea otters for meat, particularly during times when higher-calorie seals and sea lions were less abundant. According to Veronica Lech, Matthew W. Betts, and Herbert D. G. Maschner, "The relationship between the intensity of butchering sea otters and the encounter rates of other species . . . creates another proxy for understanding long-term ecosystem dynamics."[24] Study of the chemical signature of otter bones from northern British Columbia suggests that the local population suffered from a lack of dietary diversity due to a reduction in their numbers.[25] The coastlines of southern British Columbia and northern Washington State saw a stable degree of marine mammal hunting for thousands of years, which "demonstrates the capacity for sustained harvesting in antiquity."[26] Archaeological data from the Oregon coast provide evidence that sea otters "have been relatively intensively harvested during the last 1000 years."[27] Overall, several lines of information suggest that humans reduced otter numbers across much if not most of their ancient range and that this exploitation is linked to a number of biogeographic circumstances. "The magnitude of use varied spatially among coastal indigenous people," as First Nations scholars summarize, yet the relationship between sea otters and people was a deadly one for ages after migrant hunters first encountered *Enhydra*.[28] The cascading environmental effects of this hunting were not as widespread and intense as what occurred during the historic fur trade, but they can be detected.

Sea otter fur was used and traded by coastal societies prior to the interest of non-Natives in it as a commodity. Fur robes were worn by chiefs and other elites throughout the Pacific Northwest, and soft pelts were utilized as bedding. Trade networks allowed goods to flow between indigenous groups and could be monopolized to enhance the power of local leaders.[29] Sea otters likewise became integrated into the elaborate mythologies and rituals of many Pacific cultures. The Ainu of Hokkaido and the Kuril Islands, in an oral epic known as *Kutune Shirka*, tell of a hero's quest for a golden sea otter and the conflicts that resulted from his successful capture of the special creature. Animals in Ainu tradition were granted supernatural abilities and thought of as disguised gods.[30] Ryan Tucker Jones summarizes Aleut beliefs about the species: "As with the Kamchadals [Natives of the Kamchatka Peninsula], animals in the Aleut world often took on a human persona; sea otters were said to have originated as a pair of incestuous human lovers who were punished by the gods and made to inhabit the ocean. Aleut stories suggest they endowed the sea otter hunt with special significance, and killing them (and fur seals) was thought of as defeating a rival warrior."[31]

Aleuts imagined them as being involved in moral evaluations of hunters; finding his prey meant that the hunter was a worthy individual, and he was thus sought out by the otter instead of the other way around. Some scholars suggest that this proclivity to personalize the sea otter corresponded with Aleut distaste for otters as food (since eating them could be considered cannibalism) and effectively complicated hunting practices.[32] Anthropomorphism would do more to aid marine mammal management and conservation by the twentieth century, when the "cute," humanlike qualities of the sea otter became an important component in building public support for it. Yet other ancient nature traditions worked to mitigate the depletion of faunal resources. Territorial governance systems in the Pacific Northwest meant that leaders were called upon to manage plant and animal numbers. Intertribal diplomacy as mediated through gift-giving and potlatch customs helped to ensure stable supplies of marine goods.[33] Such political and economic strategies in turn conveyed the metaphysical importance of elites and their abilities to control the spirit-infused world around them.

In some Haida stories, a hunter's wife is standing in the waves cleaning a white sea otter pelt when a supernatural being emerges and steals her away to an underwater home; the hunter follows and rescues his wife. As N. A. Sloan and Lyle Dick note, "The white (or silver) sea otter is in stories from throughout the northwest coast."[34] Farther south, in warmer climates and with humans less reliant on marine mammals for fur, mythological evidence diminishes. Yet while otters do not typically appear in the spiritual worlds of southern California Natives, rare archaeological artifacts may point to such connections. The discovery of carved marine mammal effigies in the region, including those depicting the sea otter, is perhaps related to shamanistic traditions regarding the species.[35]

In conjunction with its spiritual importance to many Pacific peoples, the sea otter contended with humans in nearshore communities for thousands of years prior to the early modern era. Killer whales, sharks, and men all tried to kill the animals and helped reshape the coastal ecologies they all shared. Yet around half a millennium ago, *Enhydra lutris* began to face an existential threat unlike anything prior. Non-Native commercial markets expanded into the North Pacific, and the maritime fur trade intensified and broadened on a transoceanic scale. As this occurred, sea otters became associated with particular geographic locations where they congregated, mirroring the iconography of the species in prehistory. Steller reported that a stretch of coastline along the southern Kamchatka peninsula was known as Bobrovi Sea, from the Russian *bobry* for adult male otter. The Japanese called them *rakko*—thus the Kuril Islands, Japan's northern frontier, were Rakkoshima. It is to this area of the western Pacific we must turn, the "Sea Otter Islands," to understand how the creatures first became subject to international greed and rivalry.

1   Rakkoshima, the Sea Otter Islands

In 1770 a boat carrying some eighty Russian fur trappers (*promyshlenniki*) arrived at Urup Island, roughly south of the halfway point in the Kuril archipelago between Japan and the Kamchatka Peninsula. Russians had been visiting the island chain since early in the century, hunting animals and trading for sea otter pelts with inhabitants of the Kurils known as Ainu—"humans," in their language. Urup eventually became known for a large sea otter colony located there. Thus, the volcanic, rectangular-shaped land mass was dubbed Sea Otter Island by fur traders and was a stopping point for a number of expeditions. Violent encounters between promyshlenniki and Ainu had occurred in the Kurils prior to 1770, yet the events of that year precipitated a major response by Native warriors. According to a Japanese report, the Russians shot and killed a number of Ainu hunters following a meeting between the two groups on Urup. Outnumbered and with their pelts confiscated, the Ainu fled south along the Kurils, where chiefs at Iturup and Kunashir Islands decided to cooperate and engage the interlopers with a large force. In 1771, on board some fifty *itaomacip* (small sailed vessels), warriors set sail for Urup. Around one hundred Russians arrived at the hunting grounds this time, and as they attempted to come to shore the Ainu attacked them. Approximately ten promyshlenniki died during the assault; others scrambled through the surf and were struck by the poison arrows of Ainu archers who had climbed the sides of the vessel and shot at fleeing survivors. An additional two or three Russians were killed, and many were wounded. The Ainu had retaken Sea Otter Island.[1]

The violence at Urup was mirrored by events in the Aleutian Islands and elsewhere in the North Pacific as Siberian hunters pursued marine mammal quarry for profit and often attempted to procure tribute (*yasak*) from indigenous people by force or coercion. Examining similar actions in the Kuril Islands offers an opportunity to compare eighteenth-century Russian colonialism on localized scales. Looking carefully at sea otter hunting and trading in the western Pacific also allows us to reframe the history of the maritime fur trade more generally, since the focus of attention is most often on events in the Aleutians and to the east. The Pacific explorations of Vitus Bering, Russian expeditions to Alaska, and Spanish, British, and American efforts from Vancouver Island to California have been thoroughly documented. Oceanic histories surrounding Kamchatka, Hokkaido, and other Northeast Asian locations broaden the scope of the sea otter trade and make it possible to tell the story of *Enhydra lutris* more completely.

Inter-Asian fur markets that existed centuries before the eastward excursions of both Bering and promyshlenniki suggest the need to expand knowledge of the exchange of skins geographically and temporally. Among historians of the North American West, rethinking of the fur trade has been aided since the millennium by a growing attention to Pacific history and the development of a "Pacific World." According to David Igler, such a concept "would have made little sense" prior to eighteenth-century breakthroughs like James Cook's final expedition, which expanded and intensified European American contact with the region. Before 1700 Spain's Manila galleons set off on transoceanic ventures (inaugurated in the sixteenth century), and Europeans formed other post-Columbian economic ties with Asia and the South Pacific. Yet the Pacific remained a "Spanish lake" and was not an integrated geographic space in the same manner as was the early modern Atlantic World.[2] The people and sea otters of the ocean's northwestern portions help us to understand some of the larger forces that made such integration possible by the mid- to late eighteenth century. The first modern markets for otter pelts and Russian expansion in the western Pacific set the stage for a degree of commercial and colonial involvement that effectively put the Pacific—particularly much of its North American littoral—on world maps. Historian Robert Hellyer has argued for the importance of the Pacific's West in the forging of links with its East and "insular middle" after 1750.[3] Preceding 1750, maritime fur trading in the West unfolded in crucial and often overlooked ways.

Despite the deadly clash with Ainu in 1770–71, Russians continued to sail to Urup to exchange items such as textiles, silks, and sugar for furs even after an attempt to establish a permanent post on the island was ended by a destructive earthquake and tsunami in 1780. Metaphorically similar to the tectonic forces of the Ring of Fire, an intensification of Japanese involvement in the Kurils also disrupted the efforts of promyshlenniki to consolidate the islands commercially and geopolitically. Earlier claims by Japan to northern reaches and a need to expel the barbarians moving south toward Hokkaido encouraged the shogunate to resist Russian encroachment in the region. Among other things, this involved attempts by the late eighteenth and early nineteenth centuries to control and assimilate the Ainu traders who acted as middlemen between Japanese and Russian merchants. As Brett Walker argues, this "middle ground" of the western Pacific led to Ainu subjugation and dependency in a similar fashion to changes that affected indigenous people positioned between rival powers in North America.[4] At the same time, the international tensions of the Kuril trade helped to limit the destruction of local sea otter populations. Russians returned to settle Urup in the 1790s, but they were frustrated in part by greater Japanese investment in the southern Kurils and an inability to access resources from the Ainu. Sea Otter Island, situated along both geological and imperial fault lines, provided a different environment from the one Siberian colonists encountered at Kodiak Island around the same time (discussed in the following chapter). The difficulties that promyshlenniki had in killing them, coupled with limited Japanese interest in fur trading for much of the eighteenth century, meant that local sea otters were less exploited than the species was elsewhere in the Pacific. Otter pelts from the Kurils may have been the first to reach non-Native markets, but for various reasons the number of animals remained relatively stable there before 1800.

Commercial Networks in the Western Pacific

For much of its imperial history, China lacked forests capable of supporting sizable populations of large furbearing animals. The clearing of northern woodlands for farms left relatively little space for such species to dwell. Reverence for fur as a rare luxury item was expressed in sources as early as the Tang dynasty (roughly AD 600–1000), a mixed-blood Chinese and Central Asian ruling house. Fur was associated with "exotic" Central Asian societies, although it may have faded in importance later during less heterogeneous dynasties. The Mongols and other barbarian groups were often identified with the frontier product.[5] Sea otter fur reached China as early as the middle of the Ming dynasty (1450–1550), an imperial era marked by increased luxury consumption. Referencing Japanese scholarship, Chikashi Takahashi cites 1483 as the earliest year for the export of otter pelts to China from Japan.[6] Generally, pelts were acquired on Hokkaido (formerly Ezo) and the Kuril Islands by Ainu hunters and traders who brought them to posts on Hokkaido, a northern island discovered by Japanese explorers as early as the first millennium. Merchants then transported them south to Nagasaki for export to China through Korea. Over time, expanding the trade of marine products allowed Japan's rulers to limit the flow of silver to China.[7]

The Ainu are a people native to islands and coastal ranges in the Sea of Okhotsk. They exhibit more body hair than other East Asians, which often earned them exotic descriptions from their neighbors, and although Caucasian features once led to speculation about their origin, the Asian heritage of the Ainu has long been established. They are multiethnic descendants of ancient to medieval Jomon, Okhotsk,

FIG. 2. A Japanese woodblock from a 1799 collection shows Ainu hunters pursuing sea otters utilizing traditional weapons. Image courtesy of The Huntington Library, San Marino, California.

and Satsumon cultures and began to appear in early Japanese royal chronicles as distinct barbarians from the North.[8] Ainu hunters killed sea otters with handheld bows (*caniku*) and harpoons (*marep*) managed from boats, as depicted in a 1799 Japanese woodblock from the collection *Nihon Sankai Meisan Zue*. On land they may have used crossbow traps rigged with aconite-laced poisoned arrows, which they used to kill bear and deer.[9] Salmon fishing was central to Ainu sustenance, as was small-scale grain and vegetable crop production.

Anthropological and historical studies have emphasized the importance of long-distance Ainu trading that helped to link the Kamchatka Peninsula, Sakhalin Island, and the Kurils with Hokkaido and Japan. Permanent settlers migrated to southern Hokkaido as early as the Kamakura era (twelfth to fourteenth centuries) and laid the groundwork for an active barter exchange between the Japanese and indigenous northerners. As Kaoru Tezuka explains, "Products were brought to and from

Hokkaido in ships traveling across the Tsugaru Strait or along coastal waters. These included items not produced in Hokkaido, such as rice, salt, tobacco, cloth, *koji* ferment, and metal which the Ainu very much wanted. The Japanese were particularly eager to receive eagle feathers, which they used to fletch their arrows, and marine products."[10]

Chinese goods obtained by Ainu and other groups from the Amur River on the East Asian mainland via Sakhalin were part of these networks—this was known as the "Santan" trade, distinct from the Kuril trade. Thus it is possible that sea otter fur reached China before Japan first realized its potential as a trade item. Whatever the case regarding the first use of an otter pelt in Asia beyond those who lived near and hunted the animals, records from the sixteenth century onward reveal that Ainu tribes thrived as marine mammal hunters and traders from southern Kamchatka to Japan's northern frontier. Fur was part of an ongoing maritime exchange system that carried Japanese and Ainu goods across the northwestern Pacific, linking the Ainu with both Japan and the Russian Empire.[11]

Japanese elites were among the first to recognize the importance of otter skins as an exotic luxury item. An eighteenth-century report records that in the 1560s "pure white sea otter pelts" from the Kurils made their way to Hokkaido, and otter fur was gifted to officials in later decades. Prior to their martyrdom by the Tokugawa shogunate in the early 1600s, the Jesuit missionaries Jerónimo de Ángeles and Diego Carvalho noted the existence of the pelt trade emanating from Urup. By the eighteenth century, sea otter furs from the Kurils were believed to have special healing properties, and using them as a cushion was thought to help with poor circulation and even the effects of smallpox.[12] As historian John Stephan notes, the islands became known during the early modern era as Rakkoshima, or the "sea-otter isles," signifying the special place of the species in Japanese frontier imagination.[13]

The Kurils, a chain of fifty-six volcanic islands stretching some eight hundred miles between Japan and Kamchatka, were, like Hokkaido, exotic barbarian lands shrouded in mystery. Both appeared in inaccurate and vague forms on Japanese maps into the early eighteenth century.[14]

It was at Hokkaido that Japanese lords managed trade with Kuril Ainu. Similar to other points of exchange in the maritime fur trade, sea otter fur was one of the most expensive items traded at the island and regulated by officials. In one analysis of prices for fish oil, cloths of tree fiber, and marine mammal skins, sea otter pelts rank at over five hundred times the cost of seal skins at Hokkaido for the year 1786.[15] Perhaps the disparity had something to do with increasing competition from Russian merchants by the late eighteenth century or with an increase in the volume of fur exports to China.[16] Whatever the case, the otter was long recognized as a valuable commodity from Japan's wild North. Animals hunted and exported by the Ainu were vital sources of symbolic power for early modern Japanese elites. Hawks, trapped on the island by both Japanese specialists and Ainu hunters, were perhaps most important to the economy of Hokkaido. Similar to the possession of furs, hawks and falconry conferred a sense of wealth and prestige to lords as fantastic things from faraway lands. Some estimates from the mid-seventeenth century calculated profits from the hawk industry at or above all taxes collected from trade ships at Hokkaido.[17]

After 1700 Japanese investment in the Kuril trade expanded with the establishment of an outpost north of Hokkaido at Kunashir Island. The curious case of merchants who were shipwrecked far beyond the island may have been related to this expanding northern trade and resulted in the first Japanese individuals to bring back information on Kamchatka and Russia. The men were reportedly traveling along the coast of Honshu in 1782 and were on their way to Edo when a storm pulled them out to sea. Jean-Baptiste Barthélemy de Lesseps, who traveled to Kamchatka in 1787 with the French explorer Jean-François de Galaup, comte de La Pérouse, noted encountering nine survivors from the crew who were shipwrecked at Amchitka Island in the Aleutians and were saved by a group of Russians engaged in sea otter trading. Barthélemy de Lesseps did not entirely believe their story, but his record is unclear about whether they were engaged in trade in the Kurils or elsewhere north of Japan. He asked one of the sailors "some questions respecting the nature of the merchandise they had saved from their wreck" and was told that it consisted mostly of "cups, plates, boxes, and other commodities," some of which were sold at Kamchatka.[18] Whatever their intended destination, the merchants traveled to St. Petersburg and met Catherine the Great, and three of them returned home in 1792 as part of an expedition led by the Finnish navigator Adam Laxman, who was sent on a Russian mission to open Japan to trade. Laxman returned unsuccessful in his effort.

Whatever Japanese traders sought in return from northern territories—furs, hawks, or fish—the Ainu were dramatically affected by trade relations with Japan. Commercial contacts changed Ainu life as natural resources were increasingly viewed for their market value and less as sources of sustenance.[19] Leaders adopted Japanese goods as indicators of social and economic standing. Sake was integrated into the Ainu metaphysical universe, offered in ritual to ancestors and deities.[20] Such shifts resulted in increased dependence on trade and local environmental depletion. The overhunting of deer for skins was particularly destructive to Ainu communities on Hokkaido, a major cause of famines there in the late eighteenth century. Thus the Ainu who fought promyshlenniki at Urup in 1770 were already living in a world turned upside down by commercial and imperial contacts. Shakushain's Rebellion, a failed mid-seventeenth-century insurgency inspired by Japanese encroachment on Hokkaido, was by that point a distant memory.

Russians in the Western Pacific

Russian extension eastward to the Pacific by the early 1700s represented a distinct commercial and political challenge to the Japanese state. Due in part to the robust Kuril trade, some in Japan believed that the entire archipelago and even the Kamchatka Peninsula were within their rightful territorial claim. The appearance of foreign "red-haired devils" in these areas contested such notions and threatened to disturb lucrative fur exchanges. As the Japanese official Mogami Tokunai summarized in 1785, "The Kuril Islands belong to Japan. Sea-otter fur is the best product of Ezochi. It has been sent to Nagasaki to be sold to Chinese ships since the old days. However, in recent years, the Russians have come to collect sea-otter furs and sell them to Beijing as a Russian product. This is a shame and a serious problem for Japan."[21]

Tokunai was a member of a mission dispatched by the shogunate to explore the Kurils for colonization and trade opportunities with the Russians. Apparent in his report is that the economic concerns of the northern frontier, epitomized by the sea otter trade, were beginning to give way to national security worries by the late eighteenth century.[22] Japan's fears of Russian encroachment can be traced to developments on the Eurasian continent centuries prior. As Cossacks and entrepreneurs began to expand east across the Ural Mountains during the late 1500s, their efforts at trapping—sable, in particular—and collecting tribute contributed greatly to the tsar's royal coffers. Detrimental effects on the indigenous human and animal populations of Siberia were also part of this movement east. By the mid-seventeenth century, Russians had reached the Sea of Okhotsk and penetrated to the Amur River (imperial China's only natural northern border), raiding villages of Natives who petitioned China for assistance. It was at a frontier town in the Russian-Chinese borderlands that the Treaty of Nerchinsk was signed in 1689. Russia agreed to recognize the Amur as the official border with China and was granted trading rights in return. After 1728 commercial exchanges between the two powers were limited largely to the border town of Kiakhta.[23]

Prior to Nerchinsk, furs collected by Russians were usually sent to European markets. Following the treaty, China became a major destination for promyshlenniki. As the supply of warm Siberian animal furs ran increasingly low by the early eighteenth century and as explorers continued to advance toward the Pacific and its abundant marine mammal herds, the Kiakhta trade became an increasingly lucrative venture. In the early years of trading at Nerchinsk and Kiakhta, squirrel and ermine sold best to Chinese merchants, and Russians benefited from a number of products in addition to tea, such as silk, gold, and supplies for Siberian outposts.[24] While some scholars have argued for an overall decline in Russo-Chinese exchanges beginning in the early 1700s,[25] others emphasize expansion. Citing Russian sources, James R. Gibson contends that a number of factors account for the growing trade at Kiakhta over the course of the eighteenth century. Among them are the 1754 removal of internal Siberian customs duties and the establishment of a bank for Kiakhta merchants. According to Gibson, "From 1755 to 1760, Kyakhta's total customs duties of 1,376,000 rubles contributed just over 7 percent of Russia's gross income from all foreign trade.... [I]n the last half of the 1700s, the China trade represented about one-half (by value) of Russia's foreign trade and from three-fifths to two-thirds of its Asian trade."[26]

Hauling pelts from the Pacific coast across eastern Siberia and toward the Manchu border cut into Russian profits from the sea otter, as did British and American competition in Chinese markets at the end of the 1700s. Nevertheless, the species was an important factor in burgeoning commercial activity at Kiakhta. Peter Simon Pallas was a visitor to the trading outpost in 1772 and wrote that "to the Chinese Kamchatka sea otters, both large (dams) and medium (juveniles), are the most important and pleasing commodity."[27] Other than the dwindling sable, sea otter was the most valuable fur product that Russians exported to China in these years. According to Pallas, it sold from 90 to 140 rubles per pelt, with various species of fox going for approximately 20 to 60 rubles.

The initial advance toward the sea otter populations of Kamchatka and the Kurils was prompted by the explorations of Cossack Vladimir Atlasov, whose report was received by the expansionist tsar Peter the Great in 1701. Atlasov told of milder winters, constant volcanic activity, and Native Kamchadals who dressed in skins, ate fish, and were, for him, foul-smelling savages. Yet other details he provided piqued Peter's interest in further exploration of Kamchatka and locations to the south. These included descriptions of marine mammals, Native trade possessions from a "magnificent people" beyond the peninsula, and a representative individual named Dembei, a Japanese castaway who had been held captive at Kamchatka. (Atlasov himself later met a violent end, being hacked to death by fellow Cossacks after stealing from them.)[28] By the 1720s daring individuals encouraged by the tsar had traveled down much of the Kuril chain, encountered the Ainu, and demanded tribute. These early forays in the western Pacific ultimately laid the groundwork for Bering's expeditions east toward the Americas. Yet establishing trading relationships south of Kamchatka and laying claim to territories there were Russian motivations as well. Japan itself and the marine mammal resources within its own expanding orbit attracted promyshlenniki decades before the sea otters of Bering Island did.

Siberian fur traders of the North Pacific were essentially free peasants and bureaucrats connected to the loose governmental systems of the province. Individuals referred to as promyshlenniki were often indigenous or ethnically mixed men accustomed to arctic and subarctic lifestyles.[29] The sea otter hunting and trading they engaged in on Kamchatka and in the Kurils in the early eighteenth century has received less attention from historians than the later Commander Islands and Aleutian hunts. As noted above, Russians often referred to the sea otter as Kamchatka "beaver"; thus one influential historian may have inadvertently underestimated the commercial importance of the species in the western Pacific.[30] Nevertheless, it is possible to establish some facts regarding pursuit of the animals in the region at this time. The first Russian expedition to the northern Kurils in 1711 found no sea otters on Shumshu Island but for an island beyond noted that Natives "do hunt sea otter in January" and traded for pelts with people farther south.[31] Stepan Krasheninnikov was a student who accompanied the Second Kamchatka Expedition and whose 1755 work (published the year he died) provided the first detailed account of the peninsula. He described local hunting methods for "sea beaver" and noted how Kuril inhabitants did not always prefer such skins for use or trade:

> They [Kamchatka Natives and Cossacks] have three different ways of catching them: first, by nets placed among the sea cabbage, whither the beavers retire in the night time, or in storms. Secondly, they chance them in their boats, when the weather is calm, and kill them in the same manner they do sea lions or sea cats. The third method is upon the ice, which in the spring is driven on the coast by the east wind. . . . The Kuriles did not esteem the skins of beavers more than

those of seals or sea lions before they saw the value that the Russians put upon them; and even now they will willingly exchange a dress made of beavers' for a good one made of dogs' skins, which they think are warmer, and a better defense against the water.[32]

Krasheninnikov also noted that people from Kamchatka to Urup paid annual tribute in sable, fox, and sea otter. His writings tended to emphasize the value that natural resources had for the benefit of the Russian state.[33] His colleague on the Bering expedition, Georg Steller, likewise provided posthumous commentary on the sea otters of the region, including the first detailed description of the species, published in 1751. Unlike Krasheninnikov's work, Steller's contains a subtle but perceptible appreciation for the aesthetic qualities of the animals and a keen interest in their characteristics and behavior, informed in part by his German Pietism and Enlightenment leanings.[34] While he could evince cruelty to nature, as his descriptions of activities during the expedition's shipwreck off the coast of Kamchatka show (for example, Steller described depriving female otters of their babies for days at a time in order to hear them wail for each other and witness the anguished longing of mothers for pups), he also displayed curiosity. After time spent in close vicinity to sea otters, Steller could not help but be taken in by their charisma. "They surpass all other amphibia in play and frolicsomeness," he wrote, likewise comparing their behavior with dogs and cats, as well as with humans: "They throw the young ones into the water to teach them to swim," Steller observed, "and when tired out they bring them to shore again and kiss them just like human beings."[35] He also discussed eating the animals for survival, as well as the importance of their skins in the fur trade. Nevertheless, in *De bestiis marinis* (*The Beasts of the Sea*) Steller expressed a particular type of enchantment with and aesthetic appreciation of the sea otter that despite the charming tendencies of the species would be rare until the twentieth century. Krasheninnikov's more detached commercial attitude prevailed during the maritime fur trade.

Promyshlenniki continued to venture to the Kurils even after Steller helped to inaugurate a fur rush east of Kamchatka. Emel'ian Basov, one of the first to attempt a voyage to the Aleutians in 1743, took sea otters from the Kurils around the same time.[36] Stephan notes that "one merchant collected 118,000 roubles for a single year's (1774) sea otter catch" from the Kurils.[37] Siberian entrepreneurs seem to have focused their attention on the northern islands for much of the eighteenth century, in part out of wariness of venturing too closely to Japan and upsetting trade opportunities there. Steller's writings indicate that knowledge of islands south of the fourth Kuril island was limited as late as the 1740s and suggest that promyshlenniki had not yet reached the otter bounty surrounding Urup. He argued for greater Russian investigation and stressed the importance of developing trade with Japan as a counterbalance to the regulations of the China trade.[38] Thus the furs of the Kurils had both commercial and national strategic implications. Gradual advancement south and west of Kamchatka had important effects on the lives of locals. Extracting furs by force and other abuses committed by sailors caused northern Ainu and some Kamchadals to flee south through the archipelago and, as already seen, occasionally put up violent resistance.[39] Additionally, Russian colonization plans for the islands began to take shape. Those who promoted permanent settlement of the Kurils argued for the needs of developing agricultural bases to provision Siberian outposts, as well as the importance of facilitating commercial contacts with Japan. A series of Japanese rebuffs in the late eighteenth century and Russian inability to deal with them effectively complicated the latter aspiration.[40]

Russo-Japanese Struggle for the Kuril Trade

In 1794 Russians launched an effort to establish a permanent colony at Urup. In that year the governor of Irkutsk recommended settlement of the island, and the newly created Northern Company (later the Russian American Company), controlled by Grigorii Shelikhov, sponsored the expedition, as Shelikhov had done for the first permanent European settlement at Kodiak Island a decade earlier.[41] Similar to Kodiak, Urup was a large undertaking and not simply a venture for itinerant traders, and Shelikhov's manager overseeing the former settlement, Aleksandr Baranov, was also given authority over the Kuril colonists. Forty men and women guided by Vasilii Zvezdochetov were dispatched to Urup by 1795. Stephan summarizes the outcome:

> Zvezdochetov's party landed on the south-east (Pacific) coast of Urup in the summer of 1795 and christened the colony "Slavorossiia." Harsh natural conditions soon undermined the whole project and nearly annihilated its members. Volcanic pumice yielded little barley or oats. Exposure and starvation decimated laboriously imported livestock. Tsar Paul's granting the Russian-American Company a twenty-year commercial monopoly in the Kurils in 1799 came as faint consolation to dwindling survivors. By that time only thirteen colonists were alive. Of these some abandoned Urup for the comparative comfort of Kamchatka. Upon Zvezdochetov's death in 1805, the colony ceased to exist.[42]

Few other details exist for this attempt to found "Kurilorossiia." Colonists did engage in marine mammal hunting and some trading with Ainu, yet agricultural efforts on the island reportedly failed.[43] The demise of the Urup endeavor was accelerated by Japan's response of fortifying the adjacent island of Iturup at the turn of the nineteenth century. Under Kondo Juzo, the shogunate oversaw a concerted effort to protect the Kurils against Russian encroachment. Impoverished Ainu communities were provided for and subject to a vigorous assimilation program.[44] A small amount of secret trading with colonists on Urup was shut down, cutting off important sources of sustenance for the Russians and further deepening Ainu dependence on Japan. Natives of Iturup were also prohibited from hunting sea otters at Urup.[45] Roads, dock works, and a small fort were constructed on the island, and the shogun likewise took direct control of Hokkaido, officially relieving local Matsumae lords who had managed the territory for some two hundred years. By 1801 an envoy visited Urup and claimed it for Japan. While officials were concerned about countering other recent foreign incursions into northern waters, including a visit to Hokkaido in 1796 and 1797 by the British explorer Captain William Broughton,[46] the fear of red barbarians was central to these colonial efforts. Such anxieties were soon intensified by the aggressive actions of Russian sailors in the western Pacific.

Nikolai Rezanov, son-in-law of Shelikhov and the individual to whom the Russian American Company monopoly on the North Pacific fur trade was granted in 1799 following Shelikhov's death, was sent by Tsar Alexander I on an attempt to officially open trading relations with Japan. Reaching the Japanese coast by October 1804, Rezanov spent more than six months in isolation at Nagasaki, the location earlier offered to Russia for limited trade. Word finally came from the shogun rejecting the new request. Rezanov seethed with anger over the experience and plotted revenge while on an inspection tour of Russian America. He convinced two young lieutenants, Nikolai Khvostov and Gavril Davydov, to strike outposts on Sakhalin Island and the Kurils. In a letter justifying his private war aims in February 1806, Rezanov cited a Japanese trading post on Urup as an additional indignity suffered by the tsar's subjects. While he felt that opening Japan by force was a valid option at one moment, he thereafter rescinded his orders while docked at Okhotsk in the summer of 1806. Yet because he did so in a confusing manner and left quickly for St. Petersburg before Khvostov could confirm the new instructions, the assaults on Japanese possessions moved forward.[47]

In October 1806 Khvostov and Davydov took hostages, burned religious structures, and stole temple objects at Sakhalin, leaving behind a notice citing the refusal to trade as justification and threatening further attacks. In the following spring they assaulted Kunashir, Iturup, and Urup and returned to further plunder Sakhalin. The captains were ultimately jailed for their actions by Siberian officials at Okhotsk but by 1809 were cleared of engaging in illegal military action.[48] Japanese warriors responded by capturing Captain Vasilii Mikhailovich Golovnin, dispatched by the Russian Navy in 1811 on a geographic survey of the Sea of Okhotsk and the Kurils. Golovnin landed at Kunashir for provisions, warned of the potential ramifications of Khvostov and Davydov's activities yet apparently unaware of preparations made by locals. In response, Golovnin's men captured a high-level Japanese merchant and returned to the island in 1813 with other prisoners for repatriation. By October of that year negotiations ended the standoff. Prisoners were exchanged, and the governor of Irkutsk formally apologized for the attacks against Japan.[49]

Thus a full-scale Russo-Japanese War waited for another century, and the Kuril Islands remained an international borderland to the extent that neither imperial power could consolidate its hold over them. A de facto boundary line between Urup and Iturup Islands was offered partial recognition in Alexander I's 1821 *ukase*, a later royal declaration attempting to reassert authority over the North Pacific (discussed in a following chapter). In 1855 the Treaty of Shimoda agreed to the demarcation and placed Sakhalin under joint possession of Russia and Japan. Finally, the Treaty of St. Petersburg, signed in May 1875 and in effect until World War II, offered the entire Kuril chain to Japan in exchange for Russian control of Sakhalin.[50] While a number of issues were involved in these negotiations over the course of the nineteenth century, the sea otter trade helped shape the geopolitical debate involving the western Pacific. Russian hunters and traders had a relatively easier task incorporating Kamchatka, the northern Kurils, and the marine resources of those lands into their empire's orbit. So long as Japan established prior commercial relationships along its territorial fringes—of which furs were an important component, though they were not as central—then the advance of promyshlenniki was frustrated in ways unlike those used in the Aleutian Islands and Alaska around the same time. Russians exchanged goods with and demanded pelts from Natives, yet the deepening reliance of Ainu middlemen on trade with Japanese merchants by the eighteenth century made it difficult for them to dominate the fur trade across the Kurils. Additionally, Japan actively defended its northern territories against encroachment. The desire to keep foreign influence at bay helped ensure that islands north of Hokkaido stayed within imperial reach, however loosely.

The Environmental Legacy

It was this struggle along the territorial peripheries of rivals that effectively spared sea otter herds the degradation they experienced elsewhere in the Pacific in the eighteenth century. S. I. Kornev and S. M. Korneva estimate the number of the animals in the Kurils prior to 1700 at twenty to twenty-five thousand and note that relatively limited hunting over the course of the century kept the population near those levels. According to the authors, "In the eighteenth century, hunting pressure was not high enough to result in population declines."[51] To be sure, hunting and tribute collection on Kamchatka and among the northern Kuril Islands reduced some local herds in the first half of the 1700s. Steller reported in the 1740s that pursuit of sea otters had declined along part of the Kamchatka coast yet was picking up farther south.[52] Still, the trouble that promyshlenniki encountered as they advanced through the islands from Ainu warriors and Tokugawa officials limited environmental damage. Additionally, tempered Japanese enthusiasm for sea otter fur also helps to explain the relative abundance of the animals into the nineteenth century. For example, Japan's development of Iturup in the 1790s may have done little in the long run to reduce the island's sea otters. Englishman H. J. Snow, who hunted otters in the Kurils during the 1870s and 1880s, discussed the 1869 construction of a new colonial office at Iturup but questioned whether it had anything to do with commercial interest in skins. According to Snow, "Having visited the island of Yetorup [Iturup] in 1873, and conversed with both Ainu and Japanese there on the subject, I know that at that time little or no attention was given to hunting the otter, nor had there been for many years previously." He believed that "the long rest from molestation" accounted for both the large numbers of sea otters that he found at the island and their tame behavior toward humans.[53]

Russians returned to Urup Island in 1828. Russian American Company historian P. A. Tikhmenev, whose two-volume work was published in the 1860s, noted that fifty men were sent to the Kurils to build structures and procure sea otter skins. According to Tikhmenev, "Hunting on Urup Island brought the company more than 800,000 paper rubles worth of furs during 1828 and 1829."[54] Experienced Aleut laborers were brought to the new settlement and were working elsewhere in the Kurils by the 1830s.

Yet it was also at this time that Russian American Company officials

FIG. 3. This woodblock by nineteenth-century Japanese artist Utagawa Hiroshige III depicts Aleut hunters and Russian vessels killing sea otters in the Kuril Islands sometime in the middle of the 1800s. Some sources have incorrectly identified the hunters as Ainu. Image courtesy of Waseda University Library, Japan.

introduced conservation efforts meant to stabilize declining stocks of sea otters and fur seals across the North Pacific. To conserve the otter trade for the future, larger hunting parties concentrated in specific areas, which were then abandoned for a period of two to three years. The rotation system, in addition to other conservation methods (discussed in the following chapter), had the effect of stabilizing yearly catches after becoming official company policy in the 1830s.[55] Hence while eventual Russian success at Urup meant that Kuril sea otters began to experience noticeable declines by the mid-nineteenth century, the timing of that success helped to ensure that the population of the western Pacific remained robust enough to inspire more intensive and widespread hunting in the region later in the century.[56]

Another reason for the stability of sea otter numbers in the Kuril Islands during the eighteenth century was the inauguration of the Aleutian hunt in the aftermath of the Bering expedition. It was on ocean surfaces surrounding islands farther east that furbearing creatures existed unclaimed by any rival empire. Russian merchants quickly took advantage, sailing far away from Kamchatka and decimating Pacific marine mammals in their wake.

2   *Promyshlenniki* and Padres

Two Spanish friars, a Jesuit in New Spain and a Franciscan in Europe, penned works in the middle of the eighteenth century that warned about Russians in the Pacific. It was fifteen years after the return of the beleaguered survivors of Vitus Bering's Second Kamchatka Expedition when *Noticia de la California* was published in Madrid in 1757. Its initial author, Miguel Venegas, was a Mexican scholar tasked with writing a history of Baja California, a peninsula colonized by Jesuits at the end of the previous century. Venegas's 1730s manuscript was considered too long and was revised in the 1750s by a fellow Jesuit historian who included reports emanating from the North Pacific by that time. *Noticia* made a logical case for worrying about Russian ventures north of California: it was "natural to think" that they would continue to advance down the coast and plant a colony in Spanish territory unless steps were taken.[1] Published in Rome in 1759, José Torrubia's *I Moscoviti nella California* (The Muscovites in California) expressed a more dire tone. Torrubia served the church in the Philippines and had crossed the Pacific, and the anxieties that animated his eighty-three-page essay took on expansive geographic dimensions informed by both his scholarship and his personal experience. He linked developments in the western and eastern Pacific by noting that "from [Kamchatka] in the past the Muscovites descended to Japan" and that they were approaching California "in a similar way." Torrubia argued that Russian ships could navigate east and south and reach dangerously close to Mexico.[2] Ultimately, the padres made the case, since in 1761 the Spanish ambassador in St. Petersburg received instructions to spy for more information on Russian activities in the Pacific. By the end of the decade, the cross and the sword were planted on the coast north of Baja.

In the spirit of Torrubia, we need to cross the ocean to understand how sea otters helped shape and were affected by international competition at both corners of their range. Spanish and Russian contests in the eastern Pacific in the late eighteenth century provide parallels with the history of the Sea Otter Islands discussed in the last chapter. Fur traders advanced into borderland regions, they were viewed as a threatening presence, and attempts were made by rivals to check their movement. Yet Japan had long-standing commercial connections with their northern frontier, whereas Spain's motivation for challenging the promyshlenniki was largely territorial in nature. Colonial authorities had little interest in developing trade with indigenous people on northern coasts and for a time had limited knowledge of the profits that Russians were realizing in China with Pacific furs.[3] Moreover, the vastness of Spain's *frontera del norte* presented distinct challenges to their attempts at forestalling or removing competitors in the late eighteenth century. A huge stretch of coastline from Baja to Alaska proved too difficult to manage and contained sea otters too inviting to foreigners. By contrast, Japan effectively rebuffed encroachment in the Kuril Islands—even with relatively late fortification efforts—in part because it dealt with a much smaller and more isolated northern geography.

The geopolitical struggle in the far western portions of North America, which expanded beyond promyshlenniki and padres and involved other non-Natives before 1800, is a story that has been told before, yet it can be reexamined through environmental lenses. What did eastern Pacific imperial conflict mean for the region's sea otter populations? For the members of *Enhydra* who encountered the Russian advance to the east in the late eighteenth century, it was disastrous. Marine mammal declines pushed hunters across Aleutian and Alaskan waters in the manner that Venegas and Torrubia feared. By contrast, the establishment of Alta California resulted in limited degradation of that province's otter herds despite numerous calls to utilize them. Spanish failure to fully exploit the maritime fur trade helped make possible various incursions into coastal territories by the early nineteenth century. It also ensured that the earliest sea otter conservation efforts would be shaped foremost by Russian policies. Authorities in Alaska and elsewhere responded to environmental destruction by implementing measures by the 1830s designed to restock regional furbearers. The somewhat successful reforms deserve mention in part because they predated the more often discussed Progressive Era marine mammal statutes and treaties by close to a century. They also underscore that it was those who were the best killers of otters—Russian overlords and their North Pacific Native conscripts—who were the ablest informers of the first modern management systems. Officials in Mexican California were ill-equipped to implement hunting regulations in the same era in part because they lacked the firsthand biological experiences that would have been gained through decades of pursuing sea creatures for profit. People on Russian coasts knew sea otters best, but that knowledge came at a terrible price to nature before the nineteenth century.

The environmental contexts examined here have been encouraged by fresh contributions to Pacific maritime historiography. While earlier attempts have been made to catalog Russian voyages and fur catches of the eighteenth century, only recently has archival material been made widely available that allows scholars to more precisely gauge the damage done by hunting in the Aleutians and Alaska. Drawing upon this information and previously published lists of vessels, Ryan Tucker Jones provides a comprehensive "List of Hunting Expeditions in the North Pacific from 1742–1800" in his study of the Russian North Pacific.[4] The data that he assembled make it possible to compare sea otter decline in Alaska with evidence from California. The new records can also inform research by conservation scientists. Cal Lensink's pioneering work in the 1960s as a wildlife biologist who detailed the history of the Alaskan sea otter should be consulted along with up-to-date studies in maritime history.[5] Lastly, we should keep in mind that patterns of either decline or stability from North American locations are connected to larger global contexts. Ultimately, both Russians and Spaniards called for the killing of sea otters in the eighteenth-century Pacific East because of ongoing relationships with the Pacific West and the demands of consumers there. Jonathan Schlesinger has recently emphasized the expansion of trade over the course of the 1700s, which brought increasing volumes of natural products to eager customers in Qing China, commercial shifts that cast a wide "ecological shadow" across the ocean.[6] If the Chinese world became ever more trimmed with fur during the century, then much of that fur came from animal bodies shadowed in places like Kodiak and Baja.

The Second Kamchatka Expedition

Shortly before his death in 1725, Peter the Great commissioned the First Kamchatka Expedition, one of his many attempts at Russian exploration of and expansion to the North Pacific. He chose the Danish captain Vitus Bering to command the venture across Siberia and to the peninsula, where Bering and crew built a ship in 1728, sailed into the Bering Strait, and explored the Northeast Asian coast. Yet discovering the relationship between Asia and North America, one of the expedition's stated goals, went unfulfilled. Thus under Bering's recommendation a second effort was launched. It included a voyage by Martin Spanberg (a member of the previous expedition), who sailed from Okhotsk to Japan in 1739, stopping briefly at Honshu and Kunashir Island in the Kurils before returning. Since his Siberian administrator believed he had reached Korea and not Japan, Spanberg was ordered to repeat the voyage, which he attempted in 1742, but he was unsuccessful due to fog and disease.[7]

Bering's second try to sail east toward America began in June 1741, when the *St. Peter* and *St. Paul* left Petropavolvsk on the coast of Kamchatka. After almost three weeks, Aleksei Chirikov, captain of the *St.*

Paul, was separated from Bering's ship in a storm and fog. He managed to reach southeastern Alaska, where a number of his men disappeared after making landfall. Assuming them killed or captured, Chirikov and his officers decided to return to Kamchatka. They skirted the southern coast of Alaska and the Kodiak Islands, traded with Aleut Natives, and brought the scurvy-stricken *St. Paul* back to Avancha Bay on October 10. The *St. Peter* was still at sea and experienced a more disastrous outcome. After getting separated from Chirikov, Bering ended up at Kayak Island in the Gulf of Alaska by July. The ship's longboat was launched to explore nearby islands, water was gathered, and expedition naturalist Georg Wilhelm Steller made observations of recent human habitation. In late August, Bering named the Shumagin Islands for a sailor who died of scurvy and was buried there. He was the *St. Peter*'s first casualty.[8]

Still in the Shumagins (on the Pacific side of the Alaska Peninsula) in early September, the expedition made its first Native contact. Two kayaks approached the *St. Peter*, and gifts were exchanged. Bering's Siberian interpreter was unable to communicate with the Americans effectively. Invited by gestures to come to shore, officer Sven Waxell decided to go out in the ship's longboat with nine other men, but after about an hour of interaction and observation he was forced to order musket fire into the air when the interpreter was grabbed and detained. The following day more trading was accomplished, but with dwindling supplies and an increasingly sick crew, Bering decided to return home. Severe storms and burials at sea were common occurrences on the journey westward. Finally, by early November the crew had had enough and decided to land. Yet the land they sighted and stopped at was not the Kamchatka coast, as many onboard had hoped. They had to winter on an unknown, uninhabited island.[9]

Bering died on the island that now bears his name on December 8, 1741. For much of the winter and spring the crew of the *St. Peter* ate sea otters, which were abundant there. Steller wrote that "for more than six months it served us almost solely as our food and at the same time as medicine for the sick."[10] Men gambled with sea otter pelts to pass the time. They also dug roots and ate sea lions, and in May 1742 they killed their first adult sea cow. The large quantity of tasty meat provided by the now extinct species allowed more men to be put to work dismantling the wreck of the *St. Peter* and building a new vessel. Nourished by the island's natural resources and aided by Steller's knowledge of dietary remedies for scurvy, the survivors sailed from Bering Island in August. In nearly two weeks they made it back to port. Thirty-one sailors out of an original sixty-seven had died. The men brought hundreds of sea otter skins back from Bering Island.[11]

Fur Rush

Emel'ian Basov was one of the first to take advantage of the new information that returned with Bering's crew. Born in a small Siberian village, Basov rose to become a sergeant in Okhotsk and originally sought support for a voyage to the Kuril Islands, for which he was granted official permission in 1741. He was aided by a number of financiers who pooled their resources and commissioned the construction of a ship on Kamchatka during the winter of 1742–43. By the time the vessel was completed, a voyage to the East had been planned, made possible by a member of the Bering expedition who led the way for Basov and crew toward the Aleutians.[12] The Basov enterprise first wintered at Bering Island in 1743, hunted foxes and sea otters, and sighted the Aleutian's Near Islands during their return in 1744. They killed some twelve hundred otters during this initial venture and made a number of other successful trips to the Commander Islands. By the mid-1740s competitors were hunting in the Aleutians.

The generic name that promyshlenniki gave to their locally built ships was *shitik*, from the Russian word for "sew." Due to the lack of construction materials on Kamchatka, hand-hewn planks were literally sewn together using seal or walrus skin, and occasionally reindeer skin was used for sails.[13] While many of these boats proved sturdy enough for the rough waters of the North Pacific, a number of *shitik* wrecked, taking cargo and crew with them. Complicating matters was that many promyshlenniki lacked the necessary navigation experience needed for sailing in uncharted waters.[14] In the early years of Russian activity in the Aleutians, independent servicemen and entrepreneurs financed and crewed the vessels, with each participant being granted a particular number of shares. Kamchadal Natives (or Itelmen) were often hired as cheap labor for fewer shares. As Lydia T. Black notes, owners believed that indigenous hunters "were less prone to diseases and, because of their diet, were especially resistant to scurvy."[15] Collectively, Siberian crews provided hunting excursions with an intimate knowledge of marine mammals and other furbearers in addition to necessary fishing and ship-building expertise.

Typically, a crew first landed in the Commander Islands, gathered animal skins and provisioned there for about a year, and then moved eastward toward other islands and the Alaska mainland. Once one of the "Distant Islands" was reached, the crew established a main camp and divided into smaller parties, known as *artels*, to look for furs. Often *artels* were stationed near Aleut villages. This created the possibility for sexual exploitation of Native women and facilitated hostage taking—a Siberian practice and one that had been utilized among indigenous Alaskans prior to Russian arrival—in order to guarantee safety and encourage the collection of pelts.[16] While some scholars have emphasized the positive outcomes of Aleut-Russian relations in the eighteenth century, such as literacy and reduction of tribal warfare, violence and the threat of it were common occurrences in the subjugation of Aleutian people. The 1740s may have been relatively bloodless, but deadly encounters between promyshlenniki and island Natives occurred early on and throughout the rest of the century.[17] A 1745 expedition to the Near Islands resulted in a number of Aleuts being shot after a trading attempt went awry. According to the Reverend William Coxe, an Englishman who traveled through Russia and wrote an early history of Russian voyages in the North Pacific in 1780, the leader of a reconnaissance party in a later incident "treated the inhabitants in a hostile manner; upon which they defended themselves as well as they could with their bone-lances. This resistance gave him a pretext for firing; and accordingly he shot the whole number, amounting to fifteen men, in order to seize their wives."[18]

Such abuses resulted in violent responses against other groups of hunters. One vessel in 1761 lost a crew of twelve veterans of the Aleutian trade in an assault while at Adak Island.[19] Despite periodic resistance, many commercial outfits were ultimately successful in co-opting labor by force, while some established more cordial relationships and exchanges. The skill of Aleuts and Alutiiq (Kodiak islanders) at hunting sea otters from skin-constructed kayaks known as *baidarkas* became vital to Russian enterprises in the eastern Pacific by the end of the eighteenth century. While Itelmen traditionally pursued marine mammal quarry, they were essentially riverine people who lacked the maritime skills necessary to exploit sea otters on the scale demanded by investors, and other promyshlenniki were principally land hunters. Dwelling on a treeless environment surrounded by the bounty of the ocean, an Aleut "could rightly said to be more often on the water than any other people north of the equator," according to Jones.[20]

Due to the high prices already established in China for sea otter pelts, Russian ventures in the eastern Pacific gave particular attention to procuring them. Coxe noted this preference when he wrote that "the skins of the sea-otters are the richest and most valuable" of all those brought back to Kamchatka from the Aleutians."[21] His countryman and zoologist George Shaw, farther removed from Asian commerce but in possession of a specimen, noted the commercial importance of the species "to the Russian nobility and to the Turks, [their skins being] a principal article of their magnificent dress."[22] The illustration accompanying his description of the animal likewise reflected rudimentary European knowledge of it but rightly linked the creature to the pursuit of other marine wildlife.

According to Russian chronicler Vasilii Nikolaevich Berkh, "The price for a sea otter was 10 rubles at Iakutsk, but they were sold to the Chinese for 60 and 90 rubles."[23] Other furs outnumbered sea otter in overall volume. Various types of foxes were valued in Russia and likely represented a majority of the North Pacific fur trade.[24] Nevertheless, the difficulties inherent with exchange at Kyakhta notwithstanding, hauling sea otter pelts from the Pacific coast to Chinese border markets was the key motivator that carried scores of vessels far away from

FIG. 4. This painting accompanied British zoologist George Shaw's *Musei Leveriani Explication* in the 1790s. It appears to have been influenced in part by Georg Steller's sea otter illustration earlier in the eighteenth century. Smithsonian Libraries, Wikimedia Commons.

home during the fur rush. By the end of the eighteenth century, the search for abundant otter grounds brought Russians all the way to the Alaskan mainland, and *shitik* voyages lasting five or more years were not uncommon.

The Spanish Response

Despite attempts by officials to keep Russia's Pacific discoveries secret, news leaked out. Brief accounts of the Second Kamchatka Expedition and Bering's death appeared by the mid-1740s. An anonymous 1748 account written in both Russian and German was published.[25] The French cartographer Joseph-Nicolas Delisle, formerly of the Russian Academy of Sciences, produced two maps in France in 1752 and falsely asserted that Bering never made it beyond the island on which he perished.[26] Some Soviet historians argued that misinformation by Westerners was deliberate in order to undermine the tsar's imperial claims in the North Pacific. More likely is that state obsession with secrecy (driven by factors such as Russian cultural isolation and xenophobia) was responsible for the vague and piecemeal information regarding maritime expansion.[27] Venegas's *Noticia de la California* was published in 1757, and an English version appeared two years later. As it warned its international audience, "Is it not natural to think that the Russians in future voyages [after 1741], will come down as low as cape Blanco: and if California be abandoned by the Spaniards even as far as cape San Lucas? and we may well suppose that they who to-day take a view of the coasts and country, may to-morrow determine to plant colonies there. . . . How shall we hinder the Russians from making settlements there, unless we be beforehand with them?"[28]

Yet neither *Noticia* nor *I Moscoviti nella California* mentions the sea otter trade. Spanish officials apparently had little knowledge of the commercial benefits that promyshlenniki were reaping in the Aleutians until a diplomatic report in 1764.[29] The foggy nature of Madrid's foreign intelligence is reflected in a 1768 map produced by a judge in the Philippines and forwarded to the royal minister Manuel de Roda y Arrieta (who played a key role in the expulsion of the Jesuits from New Spain the previous year). It imagines the Aleutians extending in a straight line from Kamchatka to just seventy-five miles off the Alta California coast.[30]

Precisely where Russians were in the Pacific and what they were doing there was less important than the fact that they were simply too close for comfort.

Colonization of Alta California was more the result of broad changes in the eighteenth-century Spanish state than it was related to concrete information regarding Pacific hunting. Among numerous reforms, the first Bourbon kings of Spain sought to streamline imperial administration and ensure that colonial revenue was efficiently collected. Of major concern to Carlos III, whose reign began in 1759, was not only news

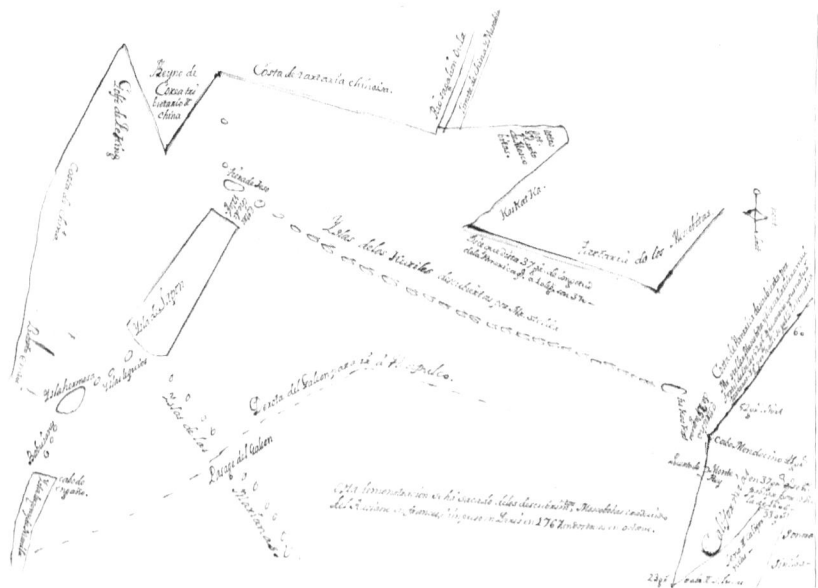

FIG. 5. Pedro Calderón y Henríquez manuscript map of the route from the Philippines to northern California, "Memorial to Don Manuel de Roda y Arrieta," April 19, 1768, MS Vault 69, California Historical Society.

of Russian encroachment but the outcome of the Seven Years' War in 1763. Spain's ally France was forced out of Canada, and the Protestant British emerged as a major North American threat.[31] With this in mind, Carlos appointed Madrid judge José de Galvez as *visitador-general* of New Spain in 1765. Galvez was given broad powers to reform and protect American holdings, and at a 1768 meeting in San Blas—a naval base city in western Mexico founded the same year—he enunciated the plan to occupy the coast north of Baja. In addition to mandating a sea expedition to Monterey, a report from the meeting noted that "it was also agreed upon that it would be most important to undertake an entry or search by land, at the proper seasons, from the missions to the north of California, so that both expeditions might unite at the same harbor of Monterey . . . for protecting the entire west coast of California and the other coasts of the southern part of this continent against any attempts by the Russians or any northern nation."[32]

By 1769 the overland party from Mexico, including the Franciscan padre Junípero Serra, who traveled north from Baja, had constructed a mission at San Diego. Later that year San Francisco Bay was discovered by land—earlier Spanish sailors had missed it in coastal fog and dangerous waters—and in 1770 a presidio and mission were established at Monterey Bay. The goal of Spanish planners was to incorporate Alta California into the empire by bringing Christianity and agriculture to coastal tribes, thus creating a "neophyte" labor force and protecting territorial claims with military outposts. The former objective met with limited success prior to mission secularization in the early nineteenth century.[33] Nevertheless, Spain did succeed in preempting Russia in establishing permanent settlements north of Mexico in the eastern Pacific. The colonization of Kodiak Island was still more than a decade away.

Such achievements notwithstanding, a series of diplomatic rumors and exaggerations in the early 1770s of planned Russian imperial expansion in the North Pacific inspired the voyage of Juan Pérez in January 1774.[34] Pérez was ordered to explore and lay claim to the coast north to sixty degrees latitude. Due to sickness onboard his ship the *Santiago*, in addition to dense fog and treacherous currents, the expedition was forced to turn around just north of the Queen Charlotte Islands (Haida Gwaii) during the summer of 1775, short of its intended goal. Yet the return trip brought Pérez and crew to Vancouver Island and a harbor Pérez named San Lorenzo: modern-day Nootka Sound.[35] They were the first Europeans to visit the location, and, while not landing, they exchanged gifts with Natives there. These actions set the stage for the later conflict with English sailors at Nootka. According to Pérez, after offering abalone shells from California, his men "got in return various sea otter skins and many sardines."[36] They were the first Europeans to trade for sea otter pelts with the First Nations of western Canada.

The mixed results of the Pérez expedition prompted a second, two-ship journey north of California in 1775 led by Bruno de Hezeta and Juan Francisco de la Bodega y Quadra. Landing in what is today Washington State, seven of Bodega's crewmen were killed by locals, and the Spanish later retaliated against an approaching canoe, incidents that were likely related to intertribal conflict in the area at the time.[37] Despite Hezeta's suggestion that the expedition return home, Bodega slipped his vessel away during a foggy night and continued north. He claimed possession of two territories near sixty degrees, including Bucareli Bay in southeastern Alaska. On his return trek, Bodega named Bodega Bay in California. Overall, with the voyages north in 1774 and 1775, Spain reached far beyond northern Mexico and preempted non-Native rivals by reaching the Pacific Northwest first. As Iris H. W. Engstrand writes, "Spain established a solid foundation for her claim of sovereignty over the Pacific Coast from Monterey to the Gulf of Alaska."[38]

Kodiak Island

The Spanish expeditions of the 1770s, including a relatively extensive return trip to Bucareli Bay in 1779,[39] momentarily bolstered the confidence of authorities that Russia's claim in the eastern Pacific was weak. That perception was soon challenged in the 1780s at Kodiak Island. The nature of the North Pacific fur trade underwent distinct changes in the last decades of the eighteenth century. Large merchant houses, capitalized from urban centers such as Moscow, began to squeeze out smaller Siberian traders. Such actions were crucial for Russian success in the trade in part because of the greater distances ships had to cover to find untapped hunting grounds. Competitors emerged that funded increasingly costly enterprises to the Aleutians and beyond and fought for commercial dominance on the coast. Irkutsk merchant Pavel Sergeevich Lebedev-Lastochkin managed expeditions to the Kuril Islands and Alaska in the 1770s. His onetime partner, Grigorii Shelikhov, sought financial and governmental support for a permanent settlement in America. Between 1782 and 1783 Shelikhov and his backers built three vessels in Siberia and launched them to sail for Kodiak, an island known to Russian hunting parties for at least twenty years.[40] Kodiak was rich in timber, densely populated with potential laborers, and replete with furbearing animals. The island was also strategically located near the mainland, where Lebedev-Lastochkin's and Shelikhov's men eventually engaged in open confrontation.[41]

Kodiak's indigenous Alutiiq (called Koniags by the Russians) had been feared by Russian hunters for their martial prowess prior to the expedition. Although official imperial policy at the time dictated that Pacific people were not to be subject to violent conquest, Shelikhov planned for a military invasion of the island. According to Black, "Perhaps he counted in advance on support from Siberian officials (whom he cultivated and bribed) should an investigation follow."[42] In August 1784 Shelikhov's vessels sailed into Three Saints Bay. Following an initial skirmish, execution of Alutiiq captives, and word of an impending assault against the colonists, the expedition's cannon were turned toward Native positions on the island and unleashed, killing some 150 or more.[43] Afterward the Russians built fortified outposts on Kodiak and the Kenai Peninsula on the Alaskan mainland. Shelikhov's somewhat self-serving account of the settlements emphasized tenderness toward locals, not slaughter. At one point he attempted to "lead them to an understanding of books" and reportedly taught a number of children Russian with some success.[44]

From Kodiak, *artels* were sent to surrounding islands, including St. Paul in the Pribilof Islands, and to Prince William Sound. As in the Aleutians, Russians relied on hostage taking to guarantee that Native men would hunt. Yet Shelikhov was aware that force and fear of retribution would not be enough to guarantee long-term stability in America. Thus he called for Kodiak islanders to be treated fairly and provided for materially. More than once he sought to have unmarried male Russian colonists marry Native women. In addition to wives, promyshlenniki often adopted Alutiiq food and dress. For Gwenn Miller, what emerged in Russian America beginning at Kodiak were curious but not entirely unique convergences of colonial "violence and dependence, both coercion and compassion."[45] The social destruction wreaked by the sea otter trade was associated with broader transformations in the lives of indigenous Alaskans.

Spanish Attempts at the Sea Otter Trade

Individuals in Mexico realized both the economic and strategic importance of the sea otter trade, and despite a general lack of interest in such ventures in New Spain they sought to capitalize on it. While padres in Baja California knew of sizable local populations of otters as early as the 1730s, and pelts were occasionally shipped on Manila galleons,[46] coordinated efforts to procure them did not begin until after knowledge of the trade expanded in the late eighteenth century. In addition to word of British captain James Cook's arrival in the eastern Pacific and dramatic news of the profits his crew made with furs taken to China, the need for Chinese quicksilver (mercury) for mines in New Spain also motivated the first fur trade plan. Vicente Vasadre y Vega gained permission in 1786 to establish a system whereby mission Indians hunted sea otters in exchange for trade items. Furs were then shipped to Canton to sell for quicksilver. By late 1786 Vasadre had collected over a thousand otter pelts from Baja and Alta California for his first trip to Asia the following year.[47]

Because California Natives by and large did not traditionally hunt sea otters, and with rumors of presidio soldiers abusing neophytes to obtain pelts for themselves, Vasadre modified his plan. Mission control of the trade was tightened, and more extensive exchange goods were sent from San Blas to the colonial capital at Monterey. Nevertheless, despite initial enthusiasm in both California and Mexico City for Vasadre's enterprise, it soon crumbled. His superiors considered the return on trade articles too low and handed over control to the military. More importantly, the Philippine Company, Spain's monopoly on the China trade, petitioned for exclusive rights to the quicksilver and sea otter markets in the mid-1780s. Both the Spanish governor at Manila and company agents at Canton frustrated Vasadre's attempts to sell his furs. In 1790 a royal decree officially ended his arrangement.[48]

Although Vasadre's otter effort was short-lived, some Spanish authorities in the Pacific realized the importance of developing the trade in order to curtail international activities on the coast. Thus while his plan was under way in 1786, the intendant of the Philippines suggested to Madrid the ambitious plan that fur trade outposts be established from California to Alaska. As he wrote, "It is undeniable that by treating those miserable natives with gentleness and affability, and by having them know the advantages of clothing themselves and of living under a pious chief, who will direct and govern them, enjoying all the rights of humanity, we shall be able to hold fast by their help a rich and powerful commerce to which other strong nations now aspire, which perhaps, if we are negligent, will gain sole rights in it when we are purposing it."[49]

The proposal was turned down by the Philippine Company as too expensive and risky, but the geopolitical realities of the otter trade took on increasing significance just prior to and following Spain's showdown with England at Nootka Sound. A series of proponents in Mexico emphasized the advantage New Spain enjoyed of already having established a series of supply bases in California, and they reminded officials of the value placed upon California abalone shells by cultures of the Pacific Northwest. Although resolution of the Nootka Controversy officially forced Spain to abandon a permanent settlement at Vancouver Island, a number of last-minute attempts were made to strengthen Spanish claims there and to deprive other nations of the commercial benefit of sea otter trading in the area. Expeditions successfully exchanged sheets of copper and abalone shells for pelts in the early 1790s. The schooners *Sutil* and *Mexicana* accomplished this at Nootka in 1792 and sailed into the Inland Passage. There they encountered the English explorer George Vancouver before returning to Mexico.[50] Despite these efforts, the Philippine Company's monopoly on Chinese goods continued to blunt attempts by merchants to engage in the otter trade either along the California coast or farther north, and the company itself abandoned its interest in it. Mercantilist policy and diplomatic concession worked together to effectively deny Spain the commercial and imperial rewards that the marine mammals of the eastern Pacific represented.[51] Failure to develop a maritime fur trade came back to haunt Spanish officials in the early nineteenth century as both Russians and Americans entered California waters to hunt. It was a consequence foretold by friar Antonio de San José Muro, who composed a long letter to Madrid in 1789 warning of the need to stop the foreign enterprises. According to Muro, Russians hunting in California was a "dire moment I see as not so distant as some others imagine."[52]

Data on pelt exports for the 1780s and 1790s confirm Spanish inability to fully exploit the marine mammal populations of the eastern Pacific. Historian Adele Ogden compiled a list of California trading vessels that allows us to roughly calculate sea otter hunting in the province (see the appendix). As Ogden noted, the information is incomplete and should be used with caution. While sources from her early research on the California trade note that Vasadre's plan resulted in 9,729 skins from Baja and Alta California, only a portion of that figure appears in the maritime records from which she compiled her 1979 vessel list.[53] Yet granting that thousands more sea otter skins should be added to the list for the years prior to 1800, her research illustrates limited Spanish pelt collection in comparison with later efforts in California. For example, for 1811 alone Ogden records 9,356 skins exported from California, the product of three American vessels and one Russian vessel hunting on the coast that year. To be sure, sea otters were slaughtered on an unprecedented scale by locals for a number of years before the end of the eighteenth century, but the killing essentially cooled off. Americans, Russians, and the North Pacific Native hunters that they both brought with them found nearshore waters cluttered with otter rafts much as the Spanish found in 1769.

From Near Miss to Muscovites in California

Despite burgeoning Russo-Spanish tensions in the late eighteenth century, the respective empires avoided an immediate border crisis in the eastern Pacific until the founding of the Ross colony in California in the early nineteenth century. The need to locate new sea otter hunting grounds and to challenge the increasing number of competitors on the Northwest coast were factors behind Russia's Billings expedition of the late 1780s and early 1790s, named for the English sailor Joseph Billings, who was then in Catherine the Great's navy. Billings and his vessels were part of a larger maritime effort to explore and protect Pacific colonies and to circumnavigate the globe, plans that were canceled in 1788 due to war with Sweden. Billings's orders were rescinded by Catherine, but he had set sail prior to receiving them.[54]

News of the Russian maneuvers reached Madrid, including false rumors of a planned settlement at Nootka Sound, recently visited by Captain Cook.[55] In 1788 two ships were ordered north under the command of Esteban José Martínez, a veteran of the Pérez expedition of 1774. His orders were to investigate Russian settlements and commerce in the North Pacific and to take possession of land when possible. However, Martínez was to avoid conflict with ships or individuals of any nation. At Kodiak and Unalaska Island to the west, the expedition learned from fur traders that while there was no colony at Nootka, a force was on the way to occupy the location and to eliminate a rumored English presence there. Thus Martínez was apparently able to gather information about the Russian circumnavigation squadron, as news of its cancellation had not yet reached Alaska. Despite cordial meetings with local promyshlenniki and the presence of a Russian outpost, Martínez performed an official ceremony of possession on Unalaska. Meanwhile, the Billings expedition was busy conducting cartographic work in the western Pacific and did not reach the island until 1789.[56]

Martínez returned to San Blas and immediately offered to proceed north again to prevent foreign occupation of Vancouver Island. Once there, he encountered English sea otter traders, and his actions contributed to a diplomatic standoff that forced Spain to abandon its exclusive prior discovery claim to the area. A direct clash between Russia and Spain was averted, but fears of Russians in California ultimately became a reality. The Russian American Company, granted a government monopoly on the fur trade of the North Pacific four years following the death of its founder, Shelikhov, in 1795, sought various ways to expand operations south of Sitka in southeastern Alaska (the capital of Russian America after 1808). Company manager Aleksandr Baranov dispatched hunting crews of Aleuts to the rich sea otter grounds of the San Francisco Bay in 1809 and 1811. Despite attempts by Spanish sentries to halt the actions of the groups, thousands of skins were obtained. Thereafter Baranov moved to establish a permanent base near the bay that would serve to supply agricultural products for needy Alaskan settlements. Fort Ross, just north of Spanish territories surrounding San Francisco, was constructed beginning in 1812.[57]

Local authorities were unable to stop the founding of the Ross settlement, but attempts to expand Russian hunting operations to the south were successfully rebuffed, and trade between Russian and Spanish colonists was limited largely to agricultural supplies. On numerous occasions, Russian American Company representatives attempted to negotiate a contract system whereby Aleuts would work deeper into Spanish waters and furs would be divided equally. Citing earlier restrictions from Madrid, California governor Pablo Vicente de Sola turned down each request.[58] Despite the restriction of Russian activity to coastal areas surrounding Fort Ross—where sea otters were quickly depleted—promyshlenniki and foreign hunters stationed in California in the 1810s represented a dramatic geopolitical turnaround for Spain. Once able to assert claims to large portions of eastern Pacific territory, the empire feuded with its original Pacific rival less than one hundred miles from the nearest presidio. Likewise, Russian imperialists achieved in the east what they were eventually able to achieve in the western Pacific, establishing a permanent colony and exploiting marine resources in intimate proximity to their borderland neighbor. Ross preceded successful settlement of Urup. To be sure, the cancellation of the sea otter trade in the late eighteenth century was less a cause of Spanish atrophy in the region than were the Napoleonic Wars of the early nineteenth century and an insurgency in Mexico beginning in 1810.[59] Yet had individuals such as Vasadre succeeded in their plans, it is unlikely that either Russians or Americans would have been as active along the California coast as they were after 1800. Once local sea otters were wiped out, the Ross colony was never more than a self-sufficient enterprise. It was finally sold to the Swiss pioneer John Sutter in 1841.

Sea Otter Decline and Stability in the Eastern Pacific

Extensive marine mammal exploitation and population decline did not precede the Russians' arrival in Alta California, but it was a main factor in getting them there. Jones charts significant decreases throughout the Aleutians and southern Alaska during the eighteenth century. Specific island groups show periods of commercial booms and busts corresponding with the discovery and near extinction of local sea otter herds. As Jones summarizes, "Russians and the imperial structures propelling and following them into the North Pacific must be accounted as the main drivers of species decline." The ecological dynamics unleashed from the west drove *artels* along Pacific coastlines and toward Spanish territories.[60]

For example, the Near Islands (just east of the Commander Islands) reached a crisis point in the early 1760s. As Jones notes:

> The reason was the same as in the Commanders: an intense rush of several vessels over the short span of several years that decimated the population before it had time to recover through natural growth rates. Added to this was the addition of Aleut hunters to the trade, which increased the speed of the depletion and allowed Russian hunters to move on to other island groups while the local Aleuts continued hunting locally. In the Near Islands, the barrage came in the years 1756–1762, when at least 8 hunting voyages sailed specifically to these islands in search of sea otters.[61]

While sea otters were not completely eliminated from the area until the nineteenth century, the boom period and subsequent crash made it necessary for promyshlenniki to continue through the Aleutians to find fresh hunting grounds. Hence the Andreanov and Fox Islands were discovered in the 1760s. Andrean Tolstykh was the first to establish relations with Andreanov islanders, and *artels* killed approximately three thousand otters during his first voyage. By the 1770s expeditions brought back fewer pelts from these central Aleutian Islands.[62] Minor attempts at conservation of sea otter herds were made by the Russian state in the eighteenth century, but *yasak* (tribute) collected by hunting outfits was a more immediate concern for royal officials. These were relatively small payments to the government by hunters, yet they helped to further deplete marine mammal populations.

The environmental onslaught drove Russians farther east and eventually to Kodiak Island in the 1780s. Shelikhov, aware of the long-term effect that human actions were having on animals in the Pacific, argued that the noticeable decrease in pelts that was putting smaller merchants out of business was one reason why approval for his trade monopoly made sense. Exploiting distant sea otter grounds continued beyond Kodiak to mainland and southeastern Alaska by the end of the eighteenth century. Prior to sending hunting parties to California, Baranov accompanied a contingency of Russian settlers and promyshlenniki to Yakutat Bay in 1795. That same year, former British navy officer and company employee James Shields sailed for Baranov to Haida Gwaii, Bucareli Bay, and Sitka Sound. He traded for otter skins with Tlingit people and noted the presence of British merchants in the area. Russian American Company officials later expressed concern that Shields had contacted his fellow countrymen and warned Baranov to "be as careful as possible in watching his actions." Yet Shields's reconnaissance efforts and his escort of a large *baidarka* hunting party to Sitka helped open the door for building a new settlement there named Novo-Arkhangel'sk in 1799.[63]

Maritime records allow for a rough comparison of the Alaska sea otter hunt of the eighteenth century with the California hunt of the first half of the nineteenth century. Comparing a graph of Jones's information on the North Pacific pelt haul (figure 6) with one based on Ogden's vessel list (figure 7) highlights the magnitude of bloodshed experienced by *Enhydra lutris* as hunters moved across Alaskan waters prior to 1800. Whereas the California hunt peaked from 1801 to 1809, when 22,578 pelts were collected and exported from the coast, the Aleutians produced a comparable number of pelts—above 20,000—as early as the 1750s, and for the rest of the eighteenth century the total catch brought back by Russian vessels never dipped below that threshold for any decade. A dramatic spike for the years 1790–97 is partly the result of a decade-long take ending in 1793 of 64,000 sea otters by Shelikhov's company. The figure is a dramatic punctuation of steady and unprecedented mass killings of sea otters over a sixty-year period.[64] It is little wonder that an observer on the Billings expedition opined that "fifteen years hence there will hardly exist any more of this species."[65] The otter populations north and west of Spanish territories witnessed greater natural losses.

The damage left in the wake of the Russian advance became

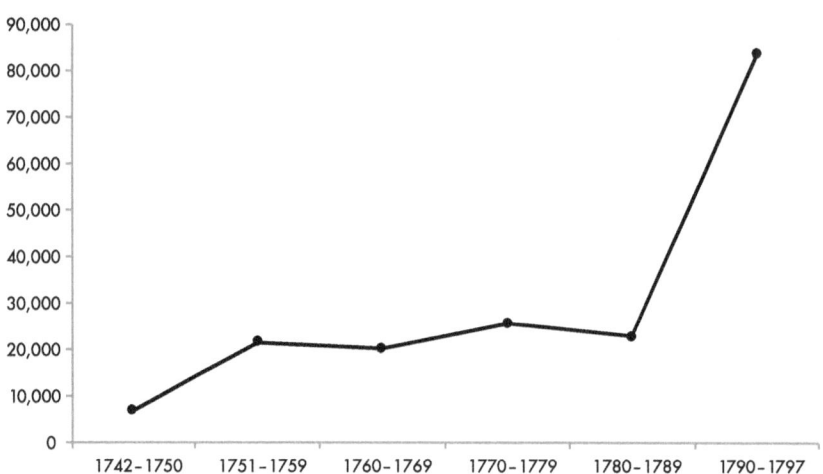

FIG. 6. A graph illustrating Russian sea otter skin exports from the Russian North Pacific from 1742 to 1797. Data compiled by historian Ryan Tucker Jones.

FIG. 7. A graph illustrating California sea otter skin exports from 1786 to 1847. Data compiled by historian Adele Ogden.

increasingly difficult for observers to miss by the early nineteenth century. The German naturalist Georg Heinrich von Langsdorff accompanied the Krusenstern expedition beginning in 1803, the first Russian around-the-world voyage and an inspection tour of Russian America. While at the recently founded Novo-Arkhangel'sk, Langsdorff summarized the historic plight of the sea otter in the Pacific:

> The decline in the number of sea otters, more apparent with each passing year, and the interest in trading in their valuable fur has caused the Russians to expand farther and farther eastward from Kamtschatka. The number of sea otters on the Aleutian Islands diminished noticeably after they began being pursued and killed in ever greater numbers. This has caused the Russians to move farther and farther, first to Kodiak, from there to Cook's River, to Prince William Sound, and then to the bays and inlets farther south and then to the Northwest Coast of America. They have killed so many sea otters of every age that they are now either almost totally extinct or have moved farther south. It is now hardly worth financing hunting parties at the northern outposts.[66]

He also mentioned that the Russian American Company presence at Sitka and Russian-led otter hunts served the purpose of denying American ships the commercial benefits of fur trading in the area. Since "all the sea otters near and far are being killed by all the imported Aleuts," little was left for American outfits that visited the coast.[67] As will be seen later, "Boston Men" proved a recurring problem for Russians in the Pacific by the early nineteenth century. At the same time, Americans and Russians cooperated to exploit California's marine mammals after 1800. Spanish uninterest allowed local otter herds to recover from minor losses that resulted from Vasadre's effort. The relative health of the California population at the end of the eighteenth century was underscored by proponents of sustaining the Spanish coastal trade. Martínez reported that "the supply of skins is plentiful, since the Californias yield many." José Bustamante y Guerra, second in command to Alejandro Malaspina's scientific expedition (1789–94), seconded his superior's commercial plan by emphasizing that San Francisco and Monterey could produce "pelts in abundance."[68] Mercantilism and inflexible monopolies blunted California opportunities and contributed to New Spain's imperial weakness. Yet such factors effectively delayed local extinctions. California sea otters entered the nineteenth century relatively stable.

The role played by Aleut and Alutiiq in Alaskan sea otter decline is a complex one. As already seen, for thousands of years indigenous people had carved out existences on North Pacific islands that resulted in overhunted animal populations. The Russian presence produced a dramatic expansion of these cultural practices as hunters were carried across island groups and to mainland territories on a scale that had never been seen before. By the turn of the nineteenth century, Native flotillas accompanied by non-Native vessels had conquered environments of the Alexander Archipelago of southeastern Alaska and south to modern-day British Columbia. In what must have been an impressive sight, some six hundred Aleuts were dispatched from Sitka in 1805 to hunt in the surrounding inland waterways, resulting in considerable damage to the area's otter numbers.[69] Russian coercion may have been necessary to cause such attacks, yet they would have been equally impossible without the impressive technique and effort exhibited by the expedition's laborers.

Hunters far away from their island homes terrorized coastal furbearers and pressured locals who relied on them. In May 1810 the Tsimshian of northern British Columbia responded violently to one such incursion that was facilitated by American ships under contract with the Russian American Company. According to Russian and American reports, well-armed groups harassed the party's support vessels just north of the Skeena River and attacked and killed some of the hunters, carrying off their severed heads as trophies. The reason for the bloodshed was the loss of sea otters from Tsimshian waters. Oral histories underscore that Natives feared the disappearance of the creatures and blamed both whites and the "Gutex," or Aleuts, "who are hunting our sea otter in their skin canoes."[70] Hence Aleut and Alutiiq not only reshaped ecosystems across the North Pacific but contributed to territorial contests in the minds of both indigenous residents and newcomers. Whether they were labeled "imported Aleuts," "Gutex," or something else, they were great otter killers, and by the time of conflict in Tsimshian lands they were being transferred to California shores to challenge Spanish claims to largely untapped marine resources.

Early Conservation

Attempts were made in the first decades of the 1800s to limit the sea otter hunt in Russian America, but, as Jones notes, efforts were sporadic and loosely enforced. (They also followed largely on the heels of fur seal conservation measures—a relationship between otters and seals in terms of conservation priorities is seen in later marine mammal history, discussed in chapter 4.) In 1825 the Russian American Company board of directors called for a stop to hunting in the Alexander Archipelago, yet fear of antagonizing Tlingit people, not conservation, was apparently the primary motivation.[71] Similar to its renewed activities in the Kuril Islands, the company began to institute more concrete measures to combat species decline and stabilize the otter hunt by the 1830s. Locations in the Aleutians that had been largely stripped of the animals saw small but steady catches in the years preceding Russia's sale of Alaska to the United States. For example, the Andreanovs saw 1,188 pelts exported between 1842 and 1861, or approximately 62 pelts per year.[72] Yet even with relatively forward-thinking environmental measures, apparently no effort was made to conserve sea otters in the vicinity of Fort Ross in California.[73] Discussed in the next chapter, Mexican officials proved unable to stop Russians or Americans from bringing numbers to low levels prior to the United States' takeover in 1848. Moreover, fragmentary knowledge regarding sea otter biology and behavior and the difficulties of implementing them for management purposes—reflected in presumably unheeded calls to ban the destructive killing of females—limited the success of Russian American Company efforts in the North Pacific.[74] During the first half of the nineteenth century sea otters may have returned to specific locales from which they had been eliminated, but in others they struggled.

The mixed record of early conservation means that it was only in a limited sense that by 1867 nearshore ecosystems from the Kurils to southeastern Alaska resembled the ones that indigenous locals shaped after thousands of years of marine mammal harvesting. The suggestion by some that Russian American colonialism left behind no detectable damage to animal populations is false. Aleutian environments at the time of the Alaska Purchase may have contained the familiar mosaics of kelp forests and barrens that prehistoric pursuit of sea otters produced, yet historic destruction had occurred. After all, promyshlenniki extinguished the last sea cow. As noted by Jones, nineteenth-century Russian naturalists were all too ready to congratulate the benevolence of imperial rule for marine mammal stabilization and for controlling the supposedly innate Native and creole tendencies to overhunt. This is despite the fact that many of the ideas reformers had about Pacific nature and animal conservation were probably borrowed from Aleuts.[75] But environmental success was evident to observers in the 1800s and was exaggerated by officials precisely because it followed a bloody century for the sea otter across a large portion of its coastal home, something perpetrated both by Russians and by coerced Native people. Local extinctions and "serial depletion" were eventually replaced by moderately effective management. Scholars will continue to sift through this complicated record of Russian hunting and conservation.[76]

3 Boston Men

William Sturgis lived a consequential American life. Born in Massachusetts near the end of the Revolutionary War in 1782, Sturgis was trained in the family business as a shipping clerk in Boston. In 1798 he was onboard the *Eliza*, bound for the sea otter trade of the Pacific Northwest. By 1809 Sturgis commanded his own vessel, named the *Atahualpa*, to China, and he successfully led his crew in a fight against pirates in the South China Sea. Thereafter he formed the lucrative Boston shipping firm Bryant and Sturgis. Following the Opium War in the 1840s, Sturgis withdrew from the China trade to invest in development projects such as railroads, although he reportedly sought to avoid risky ventures. He also became a politician, serving in the General Court of Massachusetts for over thirty years. Blending his commercial and political acumen, Sturgis was involved in border disputes and diplomatic crises regarding "Oregon Territory," as Russian, Spanish, British, and American territorial boundaries in the eastern Pacific were contested and redrawn in the first half of the nineteenth century. In 1848 his work helped lead to the adoption of the far western end of the forty-ninth parallel between the United States and Canada, an imaginary line that skirts the southern tip of Vancouver Island and divides the Strait of Juan de Fuca. Commemorating Sturgis following his death in 1863, a writer for the Massachusetts Historical Society—exaggerating somewhat—claimed that "the settlement of this dangerous controversy, by the line adopted, was mainly, if not entirely, owing to this effort of Mr. Sturgis."[1]

As one of the "Boston Men" (the generic name given to American traders by Pacific Northwest Natives) of the maritime fur trade, Sturgis was unique for acknowledging the political value of the sea otter to the extent that he did in his writings. Most of his countrymen who followed him out to the Pacific or who were later employed by him to trade on the coast were largely focused on profiting by transshipping otter skins to Canton. Nevertheless, his life and career are instructive for illustrating how American merchants and sailors dominated the otter business south of Russian Alaska by the turn of the nineteenth century and how their commercial dynamism laid important groundwork for national expansion to the Pacific Ocean. As international as the eastern Pacific was and as global as the forces were that carried Boston Men around the world time and again, understanding the American story of the maritime fur trade is crucial for understanding the history of *Enhydra lutris*. This is no simple account of national triumph and Manifest Destiny. Vessels and individuals from the United States played a central role in the decline of the species throughout much of its range up to 1850. Moreover, the exceptional accomplishments of someone like Sturgis make clear the damage that fur traders caused to indigenous societies on the coast, as he was one of the most vocal critics of the mistreatment of locals at the hands of many of his fellow merchants. Yet the United States took over a sizable portion of the sea otter's habitat at the middle of the nineteenth century, and pursuit of the animal played a part in that geopolitical change. American sovereignty further expanded in the Pacific after 1850, which led to further marine mammal population declines. The sea otter's environmental past for more than two hundred years has been defined as much by Americans as it has been by Chinese, Russians, or Aleuts.

The Americanization of the sea otter trade was energized by a series of diplomatic crises directly and indirectly related to competition for furs. International entanglements brought on by commercial activities in the Pacific Northwest influenced the political landscape of the eastern Pacific.[2] In addition to the often-mentioned Nootka Controversy, two other events are highlighted here: the struggle over the Astoria settlement surrounding the War of 1812 and the tsarist *ukase* (royal decree) of 1821. Overall, contests involving affairs in the Pacific and diplomatic resolutions in national capitals demonstrate how the maritime fur trade helped reshape North American borders. To borrow François Furstenberg's characterization of the trans-Appalachian frontier, the events suggest that a "Long War for the West" was also waged on the continent's far western portions in the late eighteenth and early nineteenth centuries.[3] While no major military engagement took place in the eastern Pacific prior to the Mexican-American War, one can trace backward from then to the 1790s and an evolving British, Spanish, American, and Russian struggle in the Northwest that was ultimately settled in the middle of the nineteenth century. That the United States emerged from the period able to successfully assert territorial claims without establishing a lasting colonial presence there is partly a testament to the accomplishments of Boston Men.

To reap the economic benefits of the trade, American merchants both competed and cooperated with agents of other nations. One example of cooperation was the licensing of Native Alaskan hunters by the Russian American Company so that U.S. vessels could effectively collect sea otters along the California coast and split profits with the company. Historians have acknowledged that "cases of conflict and collaboration often coexisted" between Russians and Americans in the Pacific during the early nineteenth century. Nevertheless, the commercial and diplomatic relationship between the two nations in regard to the Pacific was largely defined by conflict relating to sea otter enterprises into the 1820s.[4] The activities of American sailors were not only a major cause of the *ukase* of 1821 but also important in forming the United States' response to the tsar. Sturgis drew upon his intimate knowledge of Pacific Northwest trading to produce an influential essay that directly challenged Russian claims in the area.[5] As nations contended over otter habitat and marine mammal numbers were reduced up to the 1820s, new fur trade patterns increasingly linked *Enhydra lutris* with the North American beaver, *Castor canadensis*. England's Hudson's Bay Company made a concerted effort to drive American mountain men out of the Northwest interior and fortify claimed territory between the Columbia River and Russian settlements. Both American and Canadian hunters advanced into California, where sea otters were still regularly pursued with and without official license after Mexican Independence. Also, decades of visits by Pacific explorers and traders established a foreign presence at the Hawaiian Islands and radically altered the lives of Polynesian people. In broad scope, far western fur trading during the early nineteenth century linked numerous locations and facilitated human interactions in the eastern Pacific that had consequences for the ocean's humans and animals. Sustained pursuit of sea otters in the Northwest lowered their numbers to the point of the commercial irrelevance of the trade before the mid-1800s. Likewise, Russian, American, and Mexican outfits depleted marine mammal resources of the California coast prior to U.S. conquest. By that time sea otters were no longer as important to imperial expansion, but the animals would continue to suffer its consequences in an increasingly integrated Pacific World.

James Cook and the First Northwest Traders

In 1776, as independence was declared in the British colonies on the eastern seaboard, Captain James Cook set sail from Plymouth, England, on his third expedition to the Pacific Ocean. The stated intention of planners was to discover the fabled Northwest Passage, a navigable route across the northern reaches of the North American continent that supposedly linked the Atlantic and Pacific. The strategic advantage that such a waterway would afford England against its European foes was central to such geographic imagination.[6] As in his previous expeditions, Cook rounded the Cape of Good Hope and headed toward the South Pacific. In August 1777 he landed at Tahiti, and in January 1778 he "discovered" the Hawaiian Islands, where he would later be killed. By late March the *Resolution* and the *Discovery* had dropped anchor at Nootka Sound on the west coast of Vancouver Island. A robust trade with

Nuu-chah-nulth locals took place during the expedition's brief stay. As Cook's journal describes, "A great many Canoes filled with Natives were about the Ships all day, and a trade commenced betwixt us and them, which was carried on with the Strictest honisty on boath sides. Their articles were the Skins of various animals, such as Bears, Wolfs, Foxes, Dear, Rackoons, Polecats, Martins and in particular the Sea Beaver, the same as is found on the coast of Kamtchatka."[7]

After exploring the coast of southern Alaska and searching in vain for the Northwest Passage, Cook returned to winter at Hawaii, where he was killed by islanders in February 1779. The captain's men sent news of his death overland via Kamchatka, and, sailing south on their return voyage, they put in at the Portuguese colony of Macao at the mouth of the Pearl River. Sixty miles up the river at Canton, the official Manchu port open to European trade, Chinese merchants offered exorbitant prices for the crew's remaining sea otter pelts. Reportedly, the sailors were near mutiny in their desire to return to the Pacific Northwest to obtain more furs. Several crewmen even deserted in a longboat. The expedition had returned to England by early 1780.[8]

Excitement spread as official and unofficial reports of the Cook expedition were soon published. The first English businessman to send a ship to the Northwest for the purpose of the sea otter trade was John Henry Cox, stationed not in London but in Canton. Arriving in China in 1781, Cox was selling British-made "singsongs" (mechanical toys) at the time he heard of the profits made by Cook's crew. He collaborated with a number of friends in the East India Company—the English monopoly in control of the China trade—and purchased the *Harmon*, a sixty-ton vessel that was renamed the *Sea Otter* and dispatched across the Pacific in 1785.[9] Captained by James Hanna, the *Sea Otter* arrived at Nootka Sound in August. The voyage was a commercial success, gathering 560 skins and returning safely to Canton in 1786. Unfortunately, Hanna and crew left a nasty impression on the Nootka chief Maquinna. At one point the men of the *Sea Otter*, feeling threatened when Natives tried to force their way onboard in an apparent attempt to address an earlier wrong, opened fire on a canoe and killed over twenty individuals. Also, when he was invited on the vessel Maquinna was given what he was told was a seat of honor but was actually a chair with gunpowder piled underneath it. A sailor lit the powder, projecting the chief from his seat and leaving him with scars visible years later to visitors.[10] While the Nuu-chah-nulth people may have been impressed by these first experiences with explosives (which the later episode was intended to accomplish), and while Hanna did attempt to smooth over relations by tending to the injured, the actions of the *Sea Otter*'s crew came back to haunt Pacific fur traders who visited area settlements.

Other trade ventures soon followed, including a return trip to the Northwest by Hanna in 1786. With permission from the East India Company, two veterans of the Cook expeditions, Nathaniel Portlock and George Dixon, were dispatched from England. After a brief but tense stay at Hawaii during the summer of 1786, Portlock and Dixon sailed for Cook Islet and Prince William Sound in Alaska. They encountered a Russian *artel* working at the former location but still managed to obtain some sea otter furs. According to Portlock and Dixon's published account, the Queen Charlotte Islands (today's Haida Gwaii) "surpassed our most sanguine expectations, and afforded a greater quantity of furs than, perhaps, any place hitherto known."[11] Overall, they extolled the natural abundance of the coast and the prospects for bringing otter pelts to China: "On the 26th [January 1788], our principal furs, viz. the 2,552 otter; 434 cub, and 34 fox, were sold and delivered to the East India Company's Supercargoes, for 50,000 dollars."[12]

Yet British mercantile control hampered the abilities of merchants. Trade outfits were legally required to obtain licenses from the East India Company and/or the South Sea Company, the latter being the holder of exclusive rights for British trade in the Pacific Ocean. Such arrangements proved costly at the Canton market, where East India Company officials exacted sizable percentages. In addition, the company usually forbade ships from exporting Chinese goods to Europe, offering specie instead. This denied British traders the most profitable part of the exchange in the otter skin business.[13] Some attempted to bypass restrictions by sailing under falsified foreign ownership and with phony captains. One such arrangement involved Cox and former British naval officer John Mears, who along with other partners sent ships out of Macao under dummy Portuguese colors. Vessels owned by this group helped set off the Nootka Controversy, as discussed later. For the most part, however, England's merchants could not exploit the Northwest trade as successfully as their American competitors. Mirroring Spanish inabilities to pursue pelts for profit, government-sponsored monopolies held back development of maritime fur exchanges just as the United States was entering Pacific waters. Traders from the former colonies, unencumbered by complex regulatory systems, made the most out of sea otters at Canton.

American Entry

Businessmen in the United States quickly took advantage of the lucrative Northwest–Canton route. With access to British ports and the West Indies limited following the War of Independence, New England merchants sought out the China trade. Chinese luxury goods—porcelains, silk, and wood carvings—were in demand in America at the time, and tea had been a popular drink since the colonial era. Yet gold and silver, the preferred method of payment at the Canton market, was limited following the war with England. Thus traders initially tried exporting North American ginseng to China. Boston entrepreneur Sullivan Door described its all-purpose medicinal use in Asia: "It promotes digestion, procures appetite, calms the mind . . . , procures easy births, gives vigour to old and young."[14] The first American ship to make port in the Far East, the *Empress of China*, left New York in February 1784 carrying a load of ginseng and other raw materials. Sailing via the Cape of Good Hope and the Indian Ocean, the vessel returned by May 1785 laden with tea and additional trade products. The *Empress* brought its financiers a modest 20–30 percent profit on their investment.[15]

Unfortunately for the Americans, Chinese customers preferred Korean ginseng. Most traders therefore relied on either British credit for manufactured goods to trade in China or a separate exchange medium.[16] The Pacific fur trade was particularly attractive to commercial outfits in Boston, which were generally short on specie and tended to outfit smaller vessels that were more easily maneuverable throughout the many islands and waterways of the Northwest coast. As historian James R. Gibson summarizes, "The coast trade became a Boston trade."[17] The first American ships to engage in a sea otter enterprise, the *Lady Washington* and the *Columbia Rediviva*, skippered by Robert Gray and John Kendrick, respectively, left Boston Harbor in September 1787. Gray reached Nootka Sound first the following year but found the trading there light, partly due to the presence of British trader Mears and his associate, William Douglas. The Englishmen used aggressive means to obtain furs and built a temporary fort at Friendly Cove on Vancouver Island. Once Mears and Douglas were gone by October, customers flocked to Gray's and Kendrick's ships, and trading commenced.[18]

Deciding to winter at Nootka, the Americans obtained a greater number of skins in the spring of 1789 as the *Lady Washington* cruised the Queen Charlotte Islands and other locations, trading metal chisels for otter pelts. Meanwhile, British traders, including Douglas, returned to Nootka and encountered the expedition of Esteban José Martínez, sent north with orders to occupy Nootka for the Spanish Empire. While Spanish authorities had given notice to capture the Boston vessels as they proceeded to the Northwest, Martínez interpreted his own instructions liberally and left Kendrick and Gray unmolested.[19] After witnessing the beginning of the international crisis between Spain and England, the American captains separated. Kendrick remained on the coast in the *Lady Washington* to trade. Command of the *Columbia* was transferred to Gray as he crossed the Pacific to Canton and returned to Boston in August 1790, becoming the first American to circumnavigate the globe. Despite the expedition's meager financial returns, its Boston financiers sent Gray back to the Northwest after only one month ashore. In May 1792, as Gray plied the coast again, he "discovered" the river that now bears his ship's name.[20]

Nootka Controversy

Anglo-Spanish conflict occurred at Nootka in 1789 partly because of the diplomatic failings of Martínez and the British captain James Colnett, proceedings that were complicated by their conflicting missions in regard to occupying the location. After Mears returned to China with a load of furs, his associates reorganized their effort as a legal trading company, obtaining licenses from both the East India Company and the South Sea Company. Colnett was dispatched for Nootka in 1789 onboard the *Argonaut*. His goal was to construct permanent trading outposts on the coast, and for the undertaking the *Argonaut* carried twenty-nine Chinese artisans and workers.[21] As Colnett's superiors explained to him:

> In placing a Factory on the Coast of America we look to a Solid establishment, and not one that is to be abandon'd at pleasure; we authorize you to fix it at the most Convenient Station only to place your colony in Peace and Security and fully protected from the fear of the smaller Sinister Accidents; the Object of a Port of this kind is to draw the Inhabitants to it, to lay up the small Vessels in the winter season, to build, and other commercial purposes. When this point is effected different trading houses will be established and Stations that your knowledge of the Coast and its commerce point out to be the most advantageous.[22]

Given these instructions, a showdown between Colnett and Martínez—the latter having erected and christened fortifications at Friendly Cove by June 1789—was perhaps inevitable. Yet Colnett himself was partly responsible for a heated exchange onboard Martínez's vessel, *Princesa*, shortly after arriving. He falsely claimed that he was acting on behalf of the king of England in attempting to fortify Nootka and apparently insulted Martínez. The equally blustery Spanish don arrested Colnett and crew and had them sent to Mexico for trial at the end of July.[23] News of the incident had reached London and Madrid by January 1790. Talk of war soon followed, especially after the arrival of Mears, who worked to inflame Parliament regarding his losses at Nootka. Funds for a military engagement were approved while Spain sought outside support during the standoff. Yet with France mired in the beginnings of revolution and with Russia and the United States remaining neutral, the Spanish foreign minister, José Moñino y Redondo, conde de Floridablanca, was compelled to accept diplomatic terms. Spain renounced exclusive title to Nootka and admitted that the seizures by Martínez were illegal.[24] As Warren L. Cook explains, "Madrid's yielding had little to do with the validity of Spanish claims to the area in question; the consequence was a matter of which contender could marshal the most coercive power in Europe."[25]

The economic benefits of the sea otter trade figured prominently in the proceedings of British officials during the controversy. Mears was questioned in 1790 before the Committee of Trade and Plantations. He was asked repeatedly about fur trafficking and other areas of Pacific commerce such as whaling.[26] Mears's interview suggests that authorities in eighteenth-century England closely associated international commerce with national identity. The actions of Martínez were an affront to British pride not simply because of their lack of acceptable legal standing but because of economic and cartographic information that bolstered Britain's global position. For Daniel Clayton, the "ledger and the map" were powerful tools employed during the standoff over Nootka. Such nation-state traditions were "technical matrices through which Britain ordered the Pacific as a commercial arena and a space of European sovereignty."[27] Yet while mercantilist thinking aided the British in their contest with Spain, it left the empire vulnerable to competitors in the Northwest. Mears warned officials that trade there "must be carried on by one Firm or Company," and he believed that competition would lower prices too much for meaningful profit in furs. But the East India Company showed little interest in developing the sea otter trade, and Canadian fur companies concentrated for the most part on control of Rupert's Land, a territory north of the Great Lakes and east of the Rocky Mountains.[28] At a crucial moment in the history of the far western frontier, state monopoly gave way to private enterprise.

The First Nootka Convention of 1790 did not end the matter. Since occupation took precedence over claim of prior discovery during the Nootka negotiations, Spain sought to reaffirm its right to the base constructed on Vancouver Island. The diplomatic maneuver made sense due to the ambiguous wording of the treaty, which left it unclear who could settle where north of Spanish possessions in California. When George Vancouver arrived on the disputed coast in 1792 on a mission to take formal possession of Nootka (in addition to engaging in a final British quest for the Northwest Passage), he encountered naval officer Juan Francisco de la Bodega y Quadra. Cordial relations were established between the two representatives, but Bodega y Quadra refused to completely relinquish Spanish claim to the location. Maquinna, interjecting himself into the rivalry and communicating his desire for a peaceful resolution, presented sea otter skins to Vancouver and Bodega y Quadra during a ceremony in their honor. In the end, Vancouver and Bodega y Quadra could not reach a deal at Nootka. The matter was referred home for further deliberation.[29]

News of the failed negotiations reached Europe during the French Reign of Terror and a brief Anglo-Spanish alliance spurred on by the events in Paris. A final agreement on Nootka was the result. Signed in January 1794, the Convention for the Mutual Abandonment of Nootka called for the sound to be evacuated by both parties and opened only to temporary structures.[30] It was a blow from which Spanish hopes for expansion in the eastern Pacific would never recover. Despite the controversy ending with an English imperial victory in the Pacific Northwest, the extended war with France drew attention away from the possibilities of developing trade in the region. Both British mercantilism and the French wars helped pave the way for American commercial achievements in the sea otter trade at the turn of the nineteenth century.

American Domination

According to an early Massachusetts history of the sea otter trade, "At the close of the last century [1700s], with the exception of the Russian establishments on the northern part of the coast, the whole trade was in our hands."[31] A number of factors resulted in American success. Ships were outfitted with a wide variety of consumer goods for Natives. Items such as textiles, tools, muskets, and ammunition tended to be most popular. Metals, a staple of the first years of the trade, lost favor by 1800 as consumer demands on the coast shifted.[32] Vessels left eastern harbors in late summer or early fall, made the difficult trip around Cape Horn, and arrived at trading locations some six months later. Those who reached the Northwest earliest (prior to March) had the best success at turning a profit, as competition for available furs was intense. Typically, gifts were exchanged with village chiefs, and prices for skins were fixed until a captain was finally able to "break trade." Bartering took place over the side of the ship or occasionally onboard. It was in these maritime spaces that Americans and others witnessed the commercial acumen of local Indians. As the British sailor James Strange noted, "They would not part with any thing out of their hands, before that had received an equivalent; they never forgot to examine carefully our goods."[33] Traders played rival non-Native vessels off each other in order to increase prices for their skins. This practice resulted in a number of American ships colluding on the coast to fix exchange rates and divide takes equally, a countermeasure that became more common in the 1810s as sea otter numbers were falling.[34]

Not all American voyages were successful ones. As Sturgis noted, "The erroneous idea which was cherished respecting the immense profits made in the N.W. Trade induced many adventurers to engage in it without either information or Capital. The consequence was that anyone acquainted with the business might foresee, that almost all of them made losing voyages."[35] However, many Boston-Northwest-Canton trips resulted in immense returns on their initial investments. One estimation of profits is provided by Gibson: "During the 1790s an average of $62,673 worth of American trade goods were bartered on the coast for furs annually; during the same period the annual value of American shipments of sea-otter skins to Canton averaged some $350,000."[36]

The shrewdness and diligence of Yankee merchants played a key role in the Americanization of the sea otter trade. Vessels stayed on the coast, and crews gave anything they could as long as furs were available. For example, crew members on the *Hancock* in 1799 sold some of the ship's sails for "a Skin apiece," the captain sold his waistcoat and trousers for one skin, and sale of the ship's longboat garnered ten skins.[37] Captains and officers of American ships were invested in the financial success of each voyage through a system of "privilege," or cargo space for their own store of Chinese goods. They also received percentages of net profits. The advantages afforded by these arrangements were recognized by rivals in the Hudson's Bay Company: "The American Coasters are masters & part owners of their Vessels & cargoes filled out on the most economical & cheapest manner & who moreover enjoy facilities & privileges in his Canton Dealings of which British subjects are deprived."[38]

Sea otters were not the only Pacific resource that American ships brought to China as a substitute for silver. As Russians had, Boston Men carried fur seal skins in their ship cargoes, sealing at islands in the Pacific and South Atlantic where the animals congregated. As early as the 1790s, ships were carrying sandalwood from Hawaii, which was prized in China for its sweet smell and was used to make furniture and incense. The archipelago and other Pacific islands were largely stripped of the trees in the early nineteenth century.[39] Accordingly, sea otter furs made up only a portion of the overall American China trade, with specie itself accommodating up to three-quarters or more of all American imports to Canton by the early 1800s.[40] The substantial returns on investments continued to lead crews to ply the eastern Pacific for otter furs and transship them west for tea, silk, and porcelain for the voyage home. Yet *Enhydra lutris* was only one of a number of commodified species of "The World That Canton Made," a world that involved radical changes to human and nonhuman environments across an ocean.[41]

Violence in the Trade

The admonition of owners such as Sturgis notwithstanding, a number of American captains resorted to violence while in the Northwest. Fear, prejudice, and cultural misunderstanding help to explain these events. One of the most notorious American offenders was Captain Gray. At one point, anxious from earlier incidents on the coast and with a penchant for revenge, he ordered his men to open fire on a canoe "with the war Hoop" off modern-day Washington State, killing approximately twenty people.[42] Similar to their Russian predecessors in the North Pacific, several captains used hostages to obtain sea otters. Gray took a hostage at Clayoquot Sound on Vancouver Island in order to retrieve a Native Hawaiian crewmember who had deserted.[43] The transient nature of American presence in the Northwest—trading for a season or two and not settling—helps account for violent actions, as many shipmasters never returned to the region after one voyage and thus may have been more willing to use extreme measures to turn a profit. Yet the detrimental effect of aggressive tactics on commercial relations restrained others. Cruelty to indigenous people could not only upset trade opportunities but also drive customers away toward rival traders.[44]

Local leaders often responded in kind to abuses by Europeans and Americans. One of the best-known incidents of an attack against a sea otter trade vessel took place at Nootka Sound. In 1803 the crew of the *Boston* was captured, and all but two of her twenty-seven members were killed by warriors by order of Maquinna. One of those spared was John Jewitt, the ship's blacksmith, whose skill as an armorer made him a valuable captive until his rescue in 1805. Jewitt published accounts of his ordeal, helping craft his story into a popular nineteenth-century captivity narrative.[45] While the attack on the *Boston* is commonly interpreted as revenge by Maquinna for earlier insults and aggressions by non-Natives, Anya Zilberstein emphasizes the depletion of sea otters on Vancouver Island and the resulting loss of trade goods for the Native economy as a factor in the attack, since material items had earlier strengthened Maquinna's tribal position via the potlatch.[46] The killing of all but Jewitt and the ship's carpenter and the forcing of Jewitt to forge a myriad of rings, bracelets, and metal weapons for intertribal exchange suggest that environmental and material causes help explain the episode even if Zilberstein exaggerates indigenous dependence on goods obtained from traders.[47]

Thus while Sturgis lamented the "injustice, violence, and bloodshed" of the maritime fur trade, it consisted at various moments of bloody encounter begetting bloody encounter.[48] This was only one effect that the trade had on the societies of the coast. The nature and extent of change introduced by European and American commercial activities in the Pacific Northwest has been debated by historians. The "enrichment thesis" argues that exchanges with non-Natives were mutually beneficial, as the trade created wealth for many First Nation groups, as well as for Boston Men. According to Clayton, whose focus is Vancouver Island, scholars have been guilty of overgeneralizing both negative and positive impacts and have not fully appreciated the geographic variations of the trade. "Along the west coast of Vancouver Island," he writes, "wealth flowed into some Native areas and bypassed others, making some Native groups powerful and others vulnerable to colonization. And the effect of the trade on Native material culture was not egalitarian."[49] He contends that contact with outsiders intensified tribal territorial disputes but that factors like diseases introduced to local populations did not affect all people the same. A more recent call to recognize the complexity of the otter trade's effects emphasizes precontact violence in the region and how chiefs used violence to legitimize their leadership prior to the arrival of traders. By this measure Native peoples were neither bloodthirsty nor simple victims but traditionally employed aggressive actions for their own purposes.[50] Still other historians prefer more general assessments, agreeing with Sturgis that the China trade "took far more than it gave" from Pacific people.[51] What is apparent is that the sea otter trade—for better or worse—revolutionized Pacific communities by introducing them to global capitalist systems from which they had been disconnected prior to the late eighteenth century.[52]

At the Islands

Boston fur traders were the first non-Natives to visit the Marquesas Islands, some 850 miles northeast of Tahiti. The *Hope*, captained by Joseph Ingraham, stopped there in 1791. As a later merchant described the Marquesas, "The islands are broken and mountainous, the soil fertile: bread-fruit, cocoa-nuts, sugar-cane, &c., are produced here in abundance."[53] The most popular destination for provisions and shore rest during Northwest–China voyages were the Hawaiian (or Sandwich) Islands. The British captain James Hanna went there in 1785. Many a Pacific seafarer after him took advantage of the convenient location and natural abundance of the archipelago. Ships wooded, watered, and bought a variety of foodstuffs, particularly hogs. Hawaiian pigs fattened on sugarcane proved a popular island export. Gray's *Columbia* purchased 150 hogs and a number of large casks of salted pork.[54] Polynesian women gave themselves, often as a result of their husbands' encouragement.

One trader wrote that "almost every man on board took a native woman for a wife while the vessel remained, the men thinking it an honour, or for their gain, as they got many presents of iron, beads, or buttons."[55] Sexual contact was one of the main ways that disease was introduced to these Pacific communities. Charting larger infectious exchanges in the region at the turn of the nineteenth century, David Igler writes, "Endemic among European and American sailors, gonorrhea and syphilis turned epidemic when let loose among the eastern Pacific's virgin soil populations. Both diseases caused debilitating sickness and increased sterility and infant mortality rates, creating a reproductive time bomb."[56]

Partly because Hawaii proved a popular destination for merchants, whalers, and explorers alike (and missionaries arrived by the 1820s), the islands witnessed some of the more dramatic transformations of any location visited by sea otter trade vessels. In addition to disease, traders brought guns and ammunition, which were utilized by King Kamehameha to unite the islands in the 1810s. Kamehameha employed a number of British and American sailors and workmen and recruited two trusted English advisors, Isaac Davis and John Young, from otter trade ships in 1790.[57] Not only did Pacific fur traders encourage dramatic social and political shifts among Hawaiians, but they connected the islands to the broader geopolitical contests taking place in the eastern Pacific. Peter Corney was a sailor for the Montreal-based North West Company and also worked for the Hudson's Bay Company after the two companies merged in 1821. Like other "Northwesters" (a name for British fur traders of the far west), he promoted greater national involvement in transoceanic trading. After staying at Hawaii for a number of years, Corney warned London readers in 1821 that commerce between the coast, the islands, and China was dominated by Americans and Russians, "while an English flag is rarely to be seen."[58] He stressed that the Northwest "is now totally in the power of the Americans" and that American officials and merchants wished to establish overland fur outposts out to the Columbia River: "Such is the project contemplated, and if it succeed, it would have this important consequence, that it would lay the foundation of an American colony on the shores of the Pacific Ocean."[59] Corney's account was one of many sources of inspiration for the Hudson's Bay Company to make its presence felt in the eastern Pacific by the 1820s.

In California Waters

American sea otter traders also landed in Spanish California. The first visitors from the United States to California ports attempted to trade supplies for otter skins caught by mission Indians, but Spanish officials resisted what they considered contraband dealings. In response to countermeasures, to the greater difficulty of finding untapped sea otter grounds in the Northwest, and to the arrival of Russian competitors at Sitka in 1799, a number of Americans entered into a contract system with the Russian American Company to transport Native hunters to the coasts of Alta and Baja California. In 1803 Irish-born Boston sailor Joseph O'Cain first suggested the idea to company manager Aleksandr Baranov at Kodiak Island. Baranov agreed to provide laborers and *baidarkas* in exchange for an equal portion of O'Cain's take. The agreement partly fulfilled Baranov's obligation to his superiors to expand hunting operations south toward California. It also offered him a measure of control over American merchants, whose activities Russian officials were increasingly interested in curtailing. Additionally, dealing with Boston Men allowed Russian American Company skins to enter the Canton market, which was officially closed to Russia by the Chinese. Despite protests from a Spanish commander in Baja, O'Cain's expedition was a productive one. He returned to Kodiak in June 1804 with some eleven hundred furs, along with hundreds of others that he bought from California residents.[60]

Following O'Cain, more than half a dozen other captains hunted in California waters under the contract system. For example, in 1807 the *Derby*, the *Peacock*, and the *O'Cain* transported Aleuts from Russian America and were active on the coast. They steered clear of Spanish settlements, plying offshore islands and using isolated harbors like Bodega to dispatch *baidarkas* to the rich fur grounds of San Francisco Bay.[61] Overall, 1807 was the most bloody year for the California sea otter, as at least eight different vessels carried an estimated 9,784 pelts from the coast (see the appendix).

As far as can be told, only a few years in the 1790s witnessed more numerical destruction of the animals. Viewed broadly, Aleut and Kodiak islanders in the employ of Russians and Americans killed marine mammals throughout the eastern Pacific on unprecedented scales in the late eighteenth and early nineteenth centuries and played key roles in dramatically reshaping local environments. Nearshore areas from British Columbia to Baja were denuded of sea otters and fur seals through the traditional means of Alaskans. As already seen, hunters expanded their activities in the Kuril Islands in the western Pacific after 1828.[62] By that time, California otters had been killed in numbers that only their cousins along eastern coastlines suffered.

In the 1810s Russian and American hunting contracts declined. The outbreak of revolution in Mexico led to a disruption of supply ships sent to California. Padres and settlers were thus more willing than before to engage in illegal trading with Yankees for skins, helping make offshore hunting enterprises unnecessary. Equally important was the establishment of the Ross colony in California. Russian desire to strictly limit business deals with Boston traders intensified by the 1810s and helped lead to the new post north of San Francisco. As explained in an 1808 statement from the directors of the Russian American Company, the American presence in the Northwest represented both political and commercial threats to Russian America:

> The North American republicans expand their operations in places occupied by the Company and induce in the savages actions contrary to the goals of the Company. They instill among them the notion that they should not consider the Russians their oldest, most dependable and best friends, with the natural right to be their protectors not only against foreign nations but in intertribal quarrels. Such disputes have long been customary among them, and have led to their mutual destruction, with one tribe constantly warring with another over trivial insults. The republicans encourage this depravity by bringing all manner of firearms to exchange with these savages, who by their very natures and lack of education are craven and brutal.

As a result of this the savages have caused a number of unfortunate situations for the Russians who had been friendly and had had commercial relations with them.[63]

American maritime activity troubled the colonial designs of both Great Britain and Russia in the eastern Pacific. Poaching and illegal commerce in California proved mostly a nuisance for Spain compared to its larger imperial issues in the decades before Mexican Independence. Still, fur traders established commercial relations in the province that expanded during the Mexican era as a result of the "hide and tallow" trade. William Gale, one of the more prominent foreign merchants who brought American goods to California's missions and ranchos to exchange for cattle hides, was a former otter trader and employee of Bryant and Sturgis.[64] Over time, the pursuit of California's sea otter herds helped make possible the economic and political takeover of California by American nationals in the middle of the nineteenth century. Though not as publicly as Richard Henry Dana, who had alerted readers to the opportunities that far western territories provided, sailors like O'Cain had communicated similar messages years before.

Astoria

Had it not been for events surrounding the War of 1812, one specific fear of Russia and England regarding the Northwest would have been realized. The first attempt by Americans to build a permanent outpost on the Columbia River was in 1810, when the *Albatross*, owned and outfitted by Abel Winship of Boston (along with his brothers and other investors), sailed up the river. According to historians Briton C. Busch and Barry M. Gough, "The object was to have a fortified agricultural base and from there to enlarge the coastal trade."[65] Due to fear of an imminent attack from a large number of locals, the crew of the *Albatross* withdrew and returned to sailing the coast before completing a permanent structure. The wealthy New York–based merchant John Jacob Astor had somewhat better luck and more determination. Through fur-buying trips to Montreal beginning in the late 1780s, Astor learned of the early overland expeditions and Pacific fur trade plans of the North West Company. He was also inspired by reports from American otter merchants about the lucrative China trade.[66] Beginning in 1808, Astor laid out plans for a series of trading outposts along the Missouri River to the Pacific coast, roughly following the route established by the Lewis and Clark expedition a few years prior. Anticipating rivalry from Canadians, he sought official government support for his enterprise from President Thomas Jefferson. Instead of a state-sponsored monopoly, Astor secured only a statement in support of his plans.[67] Nevertheless, furthering national acquisition of territory between St. Louis and the Columbia was part of his vision for the Pacific Fur Company, even if Lewis and Clark's patron could only offer a blessing.[68]

By the spring of 1811 construction had begun on Fort Astoria on the southern bank of the mouth of the Columbia. Unfortunately for its leading proprietor, the colony was plagued by division. Astor recruited a large number of experienced British fur trappers through his connections in Montreal, which resulted in concerns regarding loyalty to the enterprise. One of his employees later recalled that "although engaged with Americans in a commercial speculation, and sailing under the flag of the United States, [the British workers] were sincerely attached to their king and the country of their birth."[69] In addition, supply and transport ships were beset with difficulties, including a disastrous Native attack on the *Tonquin* in 1811 while it was on a trading visit to Vancouver Island. The outbreak of war with England and the overland arrival of a North West Company party in late 1812 that brought news of the military conflict sealed Astoria's fate. Convinced by the Northwesters to sell their buildings and beaver pelts before the arrival of the Royal Navy, the Astorians—unaware of Astor's hurried but failed attempts in the East to protect the colony—relinquished control.

Resolution of British and American claims in the Northwest dragged on for a number of years following the Treaty of Ghent because it remained unclear if Astoria was sold as property or was captured as territory during the war. Negotiations in London in 1818 offered the temporary answer of joint occupancy of Oregon Territory. (Fort Astoria itself, renamed Fort George by the British, remained owned and operated by the North West Company.) The border between the United States and Canada was drawn at the forty-ninth parallel west of the Great Lakes, a North American demarcation agreed to by diplomats out of long-standing geopolitical assumptions and selective use of existing maps.[70] Despite its failure as a colony, by attempting to form a terrestrial linkage between the continental beaver trade and the sea otter and China trades, Astoria allowed American negotiators the opportunity to gain their nation's first—albeit partial—frontage on the Pacific Ocean. Moreover, in light of the ten-year joint-occupancy provision in the 1818 treaty, a controversial one for many Americans, it helped to shape national and international discourse on westward expansion for years afterward.[71] The commercial importance of transpacific trading remained central for many in the United States during decades of debate over Oregon Territory, even as the sea otter trade itself was in its waning years.

American Traders and the Ukase

American maritime traders were a frequent source of tension in U.S.-Russia relations in the early nineteenth century. Beyond their "illicit" (from the Russian perspective) and destructive trading of guns and alcohol to Northwest tribes, Russian American Company officials became concerned about growing colonial dependence on Boston Men. American vessels proved to be major providers of food and supplies for Russian America, as shipments to Alaska settlements from the Siberian coast did not arrive frequently enough to meet needs. Some American sailors even stayed at Sitka, including one who became an interpreter for Baranov.[72] Yet cordial and cooperative international relationships that formed in Russian America masked potent anxieties among the tsar's ministers, Russian American Company executives, and the local company manager. The first Russian consul to the United States, Andrey Dashkov, sought out Astor in the fall of 1809 in an attempt to use American merchants against each other. Since he was involved at the time in planning his settlement on the Columbia River, Astor agreed to become the sole provider of goods for Alaskan settlements in exchange for furs, cash, or bills of exchange. Astor's ships were to deposit Russian pelts at Canton and abstain from the gun trade. Dashkov and the Russian American Company believed that the deal would discourage other shipping and eliminate their American problem.[73]

Despite the fear of some officials regarding territorial rivalry on the part of Astor, the Russian government eventually formalized the agreement in early 1812. According to Howard Kushner, it was a weak move diplomatically, making colonial settlements even more reliant on the United States for supplies and transportation. Additionally, the agreement gave "de facto recognition to future American territorial claims in the Pacific Northwest" by recognizing the legitimacy of Astoria.[74] Astor's larger plan for the Pacific Northwest soon thereafter crumbled, and with it died the trade deal, leaving the Russian American Company with a lingering Yankee problem in the late 1810s. Between 1816 and 1821 over ninety American ships came to Sitka to trade, although this was a decline from previous years due to recent provisioning from Fort Ross in California.[75] Also, spurred on by the 1818 treaty with Great Britain, individuals in the U.S. Congress—most notably, Senator Thomas Hart Benton from Missouri and Representative John Floyd of Virginia—took up the cause of national involvement in Northwest fur trading, which had important consequences for U.S.-Russia relations. In December 1820 a committee chaired by Floyd offered details for a bill to occupy the Columbia River in order to secure the "Asiatic trade." The House never voted on the measure, but the committee report accompanying it contained harsh anti-Russian rhetoric, claiming that the tsar "menaces the Turk, the Persian, the Japanese, and Chinese, [and] even the King of Spain's dominions in North America" and was thus a threat to American interests in the Pacific.[76] American officials were coupling the sea otter and beaver trades and employing increasingly aggressive political blustering. Northwest furs and empire were linked in ways Astor had been imagining a decade earlier.

Benton wrote that with the Floyd committee's work, "public attention was awakened, and the geographical, historical, and statistical facts set forth in the report, made a lodgment in the public mind which promised eventual favorable consideration."[77] If congressional representatives stirred attention in the United States regarding the prospects of a permanent American presence on the coast, they strengthened Russian resolve to protect their North American territories. The Russian American Company, its original charter soon to expire, removed Baranov in 1818 and placed Russian America under naval authority.[78] The renewed company charter in September 1821 came with a dramatic statement of tsarist sovereignty in the Pacific. Alexander I declared that territory from the Bering Strait to fifty-one degrees north latitude (the northern tip of Vancouver Island) was exclusively Russian and that foreign vessels were prohibited within "100 Italian miles" (about 92 miles) from shore. Ships would be subject to confiscation within the boundary. Whether an offensive or defensive maneuver strategically,[79] American sea otter traders were at the center of motivations for the *ukase*, a fact underscored by diplomatic correspondence. The British ambassador in St. Petersburg assured London that the *ukase* was aimed at the "commerce interlope" of Americans who came to Russian America "for the purpose of interfering in [Russia's] trade with China in the lucrative articles of sea-otter skins."[80]

Then Secretary of State John Quincy Adams sought to calm the tensions between capitals and seek a negotiation. Adams was a firm supporter of American commercial interests and expansion in the Pacific and thus was interested in involving the opinions of merchants, both otter traders and whalers, in a resolution of the controversy.[81] Into the arena stepped Sturgis, who wrote a series of letters to the *Boston Daily Advertiser* in early 1822 under the pen name Circumnavigator. In them he was critical of Floyd and his supporters on various grounds, including the notion of an American settlement on the Columbia River, which Sturgis felt was too impractical based on knowledge gained from explorers and traders. He argued instead for a settlement on the Strait of Juan de Fuca even as he expressed skepticism regarding the value of Oregon Territory for U.S. commerce. According to Sturgis, "The western shores of North America produce nothing valuable for export (furs only excepted) but what is common to both sides of the stony mountains"; yet in a brief history of the sea otter trade he noted that the only trade "worth pursuing" was north of fifty-one degrees latitude, precisely into territory deemed to be Russian in the *ukase*.[82]

By October 1822 Sturgis had moved more directly into the fray with an article in the *North American Review*, an essay that had a distinct influence on the U.S.-Russia negotiations.[83] He critiqued Russian imperial claims by casting doubt on the southern limits demanded by the tsar. Conceding that Russia has a "plausible foundation" to areas encompassing the Aleutian Islands, Cook's River, and Prince William Sound, he sought to weaken any title to southeastern Alaska and the rest of the Northwest in part by noting that Spanish explorers could make stronger arguments for first discovery.[84] The history of the maritime fur trade and Sturgis's own experiences were marshaled in an effort to protect American interests in the region. If Russia's claim to Sitka could be undermined, since it was a colony established in 1799 only after English, American, and other traders had ventured to that part of the coast, then its leaders had no right to forbid foreign shipping to places where a lucrative trip could still be made. He contended:

> It is well known to the Russian fur company, that nearly all the sea otter skins, and most of the other valuable furs, are procured north of the 51st degree, and if "foreign adventurers" can be prevented from approaching that part of the coast, the company would soon be left in undisturbed possession of the whole trade, for south of 51° it is not of sufficient value to attract a single vessel in a season. This would not only secure to them a monopoly in the purchase, but give them the control of the Chinese market, for the most valuable furs, which would be still more important.[85]

For readers not concerned with a Russian monopoly on furs, Sturgis pressed worries regarding territorial advancement farther south along the coast. He warned of the settlement at Ross and of a potential conquest of California, which would place a "formidable population" of the tsar's subjects west of American boundaries. Overall, the *North American Review* article helped buttress an uncompromising diplomatic stance against the *ukase*. Secretary of State Adams demanded not only that

American ships be free to trade but that Russia restrict its aggressive claims. The final result was a treaty signed in 1824 in which Americans were allowed to trade freely for ten years at any unsettled point along the coast. Russia agreed not to form any new colony south of Sitka. Trading in guns, ammunition, and liquor was officially banned, but local authorities were not allowed to enforce the prohibition.[86]

The Columbia Department

The year following the treaty, the Russian government signed a similar accord with Great Britain, guaranteeing the right of British citizens to trade in the Northwest.[87] The agreement coincided with a renewed attempt by Canadian merchants to deny both Americans and the Russian American Company the benefits of the region's fur exchanges. In 1821 the North West Company merged with the Hudson's Bay Company, creating a new, well-financed company directed from London.[88] Retaining the name Hudson's Bay Company, it reorganized its Columbia Department—the name of its Oregon Territory district—under the leadership of George Simpson in order to secure the trade against competitors. While Americans were already turning to other opportunities in the Pacific such as sandalwood, ships still visited the Northwest coast, and crews were increasingly obtaining beaver pelts. As Gibson notes, "By the early 1820s American Nor'westermen were taking from 3,000 to 5,000 beaver pelts to Canton from the Northwest Coast annually."[89] As land furs were being siphoned away from British posts in western North America, Simpson oversaw a hurried expansion of interior and coastal operations and an intensification of competition for pelts. Forts Vancouver, Langley, Victoria, and Simpson were among the Hudson's Bay Company's numerous accomplishments across the region. Outbid and outmaneuvered for more than a decade, Americans largely abandoned coastal trading before 1840.[90]

For Hudson's Bay Company officials, one way of quelling American activity on the coast was to eliminate the remaining sea otters from Northwest nearshore environments. Simpson wrote in 1847 that while the species was becoming more numerous in the Aleutian Islands and other Russian possessions, "to the south of the parallel of sixty degrees, they have become pretty extinct."[91] This was partly the result of the collection of hundreds of pelts by the company in the mid-1830s. Otters still existed in sizable numbers at the time along the coast of modern-day Washington State, and efforts by British outfits reduced the local population prior to U.S. sovereignty, particularly in the area of Cape Flattery. While the company attempted to conserve beaver and other furbearers in overhunted areas of western North America, the sea otter was not given the same consideration.[92] In 1837 the Hudson's Bay Company subcontracted with John Bancroft, who was employed in the sea otter trade by an American firm at the same time. Bancroft was one of a handful of captains who began recruiting Natives from Kaigani in southeastern Alaska in order to hunt in the Northwest and in California. He paid what he owed the company in pelts and had some success for his American employers as well until 1839, when Native hunters onboard the *Lama* mutinied while off the California coast, killing Bancroft and seriously wounding his wife, who later died.[93] Ultimately, the Hudson's Bay Company had limited success in hunting and trading south of the Columbia River, but officials were eager to curtail American commercial activity all along the far western coast and in inland Oregon Territory, going so far as to enact a damaging "fur desert" policy in the Snake River country (modern-day southern Idaho) by intentionally overhunting beaver.[94]

A political result of these environmental measures in the Columbia Department was the inability of the United States to successfully assert territorial claims north of the forty-ninth parallel. American migrants who crossed the Oregon Trail, and the renewed border dispute they helped inspire in the 1840s, allowed U.S. diplomats to strengthen their case for lands south of the boundary. Still, even as Sturgis admitted in another influential work in 1845 regarding the northwestern borderlands, Great Britain had rights that needed to be recognized, and his countrymen who called for the "whole of Oregon" were shortsighted. While Sturgis was not the first to suggest extending the existing border line between Canada and the United States farther westward, around the southern edge of Vancouver Island and out to the Pacific, he had important readers in government circles.[95] Sturgis may have played a role in drawing the line on the map, but other Boston Men contributed to the long-running Oregon Territory debates before the 1840s by offering American leaders an oceanic view of the value of Northwest territory. When an Ohio congressman commented that "we want Oregon to protect our fisheries and our trade with China, and to put a stop to the unscrupulous sins of Great Britain," he was expressing sentiments shaped not only by heightened international tension but also by decades of visits to the Pacific by otter traders.[96]

Sea Otter Decline in Mexican California

Merchants from the United States were the main drivers of sea otter decline in the Pacific Northwest in the early nineteenth century, and Americans had the greatest impact on the animals after most of the remaining non-Russian hunt moved south to California. After Mexico officially opened California to foreign commerce in 1822, a number of factors worked toward making this otter population reduction possible, one that may have been preceded by a degree of population recovery following the close of Russian and American contract hunts ten years earlier.[97] Mexican authorities offered Russian crews limited contracts to hunt south of Fort Ross, but these were ended by 1830 as provincial leaders sought to tighten control of coastal commerce. Those who wished to settle in California and become naturalized citizens had easier access to pelts. Chief among such individuals was the English-born Boston trader John Rogers Cooper, who sold his vessel to acting governor Luis Antonio Arguello in 1823 and became a leading merchant in Monterey. Cooper—along with other naturalized businessmen like Nathan Spear—operated as an exporter of otter furs and cattle hides for American commercial agents based in Hawaii. On occasion he utilized bands of Kodiak islanders to hunt along the coast.[98] Others hunted under licenses granted to California citizens, such as former beaver trappers George Yount and George Nidever and African American hunter Allen Light, a deserter from a trade ship in 1835 and nicknamed Black Steward by his partners. Using rifles to kill sea otters from boats and Hawaiian Natives (Kanakas) to swim after and collect carcasses, Nidever and Light embarked together on various trips to the Channel Islands and elsewhere on the coast between 1835 and 1836. According to Adele Ogden, "From fifty to sixty skins were the usual returns for a trip of three or more months."[99]

The Kentucky-born mountain man Job Dye came to California in 1832 and immediately set out on sea otter hunts, including one from Santa Barbara under the license of Don Roberto Pardo. Years later Dye described collecting eighteen skins as a "successful hunt" but noted that Pardo sued him over expenses incurred, even though they had reportedly agreed to "each bearing half of the expense." Dye was thus forced to pay expenses that ate all of his profit.[100] Another contract the same year to hunt for a padre at the San Luis Obispo mission proved more lucrative. All equipment was provided along with three Native Californians as assistants. Dye collected otter skins totaling about $2,000.[101]

Provincial officials attempted to conserve sea otters by banning the killing of pups, a policy reflecting the rudimentary information in California regarding marine mammal biology. Some reformers in the Russian North Pacific made a similar call to preserve young otters, but over time others understood the futility of such a regulation if mothers were not preserved as well.[102] With less intimate knowledge of the furbearing species of their domain, California governors were ill prepared to manage hunting in this fashion, and the policy likely resulted in an inordinate number of orphaned and doomed pups squealing in the surf.[103] Thus despite licensing and regulation, sea otter numbers were substantially reduced by the 1830s, aided by the presence of several American vessels whose captains (including Bancroft, noted above) carried Northwest Indians and hunted illegally in Mexican waters. The environmental effect of these combined pressures can be estimated from Ogden's list of trade vessels (see the appendix). While totals of otter skin shipments must be read with caution, since Ogden herself noted that some ships stopped in California carrying cargoes of skins obtained farther north, a marked decline is evident. For example, the Hawaii-based American vessel the *Waverly* took 138 pelts from the coast during an 1826–27 trading visit. Hunters onboard the Russian brig *Baikal* killed 468 of the animals in the previous season of 1825. Available totals after the 1830s are more meager. The *Diamond*, a British ship, gathered forty-four pelts in 1843, and the *Barnstable* from the United States took eighty between 1842 and 1844. Sea otters could still be found late into the Mexican era, particularly in areas surrounding the Channel Islands and along the Baja coast, but they were an increasingly rare marine resource in the years leading up to the gold rush.

John Woodhouse Audubon, the youngest son of famed naturalist John James Audubon, provided a fitting tribute to the California sea otter just after the close of Mexican sovereignty. He came to California in October 1849 with a group of miners, but he hoped to continue his and his father's work in collecting natural history specimens and illustrating western nature. Unfortunately for Audubon, the trip proved a bust both in terms of striking it rich and for plant and animal collecting. Yet he did produce one finished painting that was later included in *The Viviparous Quadrupeds of North America*, published (after his father died) in the series' third volume in 1854.[104] It depicts a sea otter that Audubon reportedly saw while traveling the San Joaquin River on his way to the gold fields. The animal was shot at on two occasions by members of the party before disappearing toward the opposite riverbank. If it truly was *Enhydra* that the men encountered, it was likely one of the last individuals of the species to inhabit the inland waters and tributaries of the San Francisco Bay. Audubon's published text and the volume's sea otter plate both lament the lack of detailed information on the creatures. *Quadrupeds* reports: "In the accounts of this species given by various authors we find little respecting its habits, and it is very much to be regretted that so remarkable an animal should be yet without a full 'biography.'"[105] The painting of "*Enhydra marina*" is undoubtedly the most beautifully textured and lifelike image of the sea otter to that time, silky, glistening fur and all, but it displays the animal uncharacteristically clutching a fish. The work reflects the often tragic blend of human fascination with and lack of data on sea otters that historically plagues them.

In the wake of the gold rush, American citizens continued to pursue the marine wildlife that first called them to the Pacific. By the 1860s and

FIG. 8. This painting of *"Enhydra marina"* accompanied John Woodhouse Audubon's *The Viviparous Quadrupeds of North America* in 1854. It was inspired by a sea otter that he and his companions reportedly encountered during the California gold rush. Biodiversity Heritage Library, Wikimedia Commons.

1870s political and commercial changes in the region had shifted the sea otter hunt into an ocean-wide quest to find and kill some of the last members of the species. This story of near extermination is less studied than the prior years of the maritime fur trade, but it is the focus of the next chapter. The environmental nadir of the late nineteenth century was made possible by over half a century of seafaring activities that informed Americans of the commercial benefits of the Pacific's animals. Trading and hunting reduced populations along the western coast of North America and buttressed expansionist ambitions. Volume declined, yet memory of profits remained. Once Americans encountered sea otter grounds along shores farther to the north and west of the Golden State, creatures suffered on a scope not seen since the eighteenth century and the heyday of the promyshlenniki.

4   Near Extinction and Reemergence

In 1868 the schooner *Cygnet*, captained by Martin Morse Kimberly out of Santa Barbara, California, sailed on a sea otter hunt to the Channel Islands and the Baja California coast, locations where its crew collected about one hundred skins. Four years later, the ship was at the island of Iturup in the Kuril archipelago and carried some two hundred otter pelts to Hokkaido, Japan. It was a marine mammal haul that set off no small amount of transpacific activity once word reached back to North America. Kimberly's wife later recalled that news of the substantial hunt in the Kurils "created a good deal of excitement" in Santa Barbara as vessels followed the *Cygnet*'s wake to East Asian waters. In 1874 Kimberly was taking fur seals at the Pribilof Islands in the Bering Sea, illegal actions at the time in light of a hunting monopoly granted by the U.S. government to a San Francisco–based firm for sealing at the Pribilof rookeries. Treasury agents stationed at the islands confronted Kimberly and crew, but the captain explained that they were only interested in hunting sea otters and had some two hundred pelts onboard. Kimberly was ordered to return a few dozen fur seal skins from the *Cygnet* but was presumably left alone to kill otters. Reports by American officials in subsequent years noted that the *Cygnet* had refit in Victoria, British Columbia, and that it was again engaged in illicit seal hunting at the Pribilofs. Thereafter, Kimberly reportedly returned to the Asian coast and was lost at sea in a typhoon in 1878.[1]

The exploits of Kimberly and other men who lived and died on the *Cygnet* reveal telling details regarding the killing of Pacific marine mammals in the late nineteenth century. The fates of fur seals and sea otters were dramatically linked as maritime hunters sought out both creatures across vast seascapes and reduced their numbers to historic lows. Kimberly's run-in with federal officials demonstrates the importance of political and commercial events that shaped the late nineteenth-century hunt, including the United States' purchase of Alaska, and the focus of authorities on fur seal remains exposes the limitations of early U.S. management policies. As American ships plied the Pacific in search of the last remaining otter herds, concern regarding the environmental status of the species was muted. Both land- and sea-based hunters brought sea otters close to complete extinction as the animals were eliminated from significant portions of nearshore habitat by the early twentieth century. At the same time, U.S. conservationists began to express more vocal support for marine mammals, and the sea otter was ultimately included in an international statute, the North Pacific Fur Seal Convention of 1911. While it may have been too late to save the species from the depredations of hunters like Kimberly, Progressive Era measures were an important first step in helping it to recover in the twentieth century.

This period of sea otter near extinction and early twentieth-century preservation has not often been examined by historians, but it has been examined by biologists like Glenn VanBlaricom.[2] His and others' assessments of the efficacy of Progressive Era marine mammal policies are synthesized here. While the animals were still killed opportunistically by hunters in the first half of the twentieth century, sea otters were in the process of recovery when they were famously "rediscovered" along the California coast in 1938. The quiet existence of a raft in the Monterey Bay area was widely publicized at the time, and it has been subject to exaggeration in natural history accounts ever since. Biological expertise on sea otter history and a careful retelling of the animal's reemergence speak to the need for scholars to engage in more dialogue on these topics. If Christine Keiner is correct that "no discipline is an island" in the realm of oceanic history, then we should continue to seek deeper understandings across disciplines that will allow for richer, more complete narratives of the Pacific's faunal past.[3] This ocean-wide view of the sea otter from roughly 1850 to 1950 is one attempt at fulfilling that mission. Ironically, avoiding insularity returns us to islands. It was at the Commander Islands in the western Pacific where Russian researchers made some of the first advances in otter husbandry prior to the Monterey rediscovery. These early achievements in marine mammal conservation science are noted here but require closer attention from scholars.

Shifting Tides for Pelts

By the second half of the nineteenth century, most sea otter pelts were shipped to the eastern United States and to Europe. This shift away from Asian markets was related to the increasing trade in opium. The British East India Company and others profited handsomely from a rise in imports of the drug into China during the early 1800s. English and American traders (the latter largely dealing in Turkish opium) were increasingly able to purchase teas, silks, and other desired goods with opium cargoes as Chinese demand for the drug increased. Despite Manchu decrees attempting to ban the pernicious trade, it flourished. According to Arrell Morgan Gibson, "Yankee ship captains in 1805 delivered 102 chests of opium to Canton; by the close of the 1820s, this had increased to nearly 1500 chests a year, valued at over $2 million, representing one-tenth of American imports to China."[4]

Strengthening Chinese protest over the damaging economic and social effects of opium addiction and the British counterresponse led to the first Opium War beginning in 1839. In August 1842 the Treaty of Nanking officially ended the conflict, ceded Hong Kong to Great Britain, and opened an additional four ports to foreign commerce. A

FIG. 9. A painting of the schooner *Cygnet* by Frank Wildes Thompson. Courtesy of Santa Cruz Island Foundation.

Sino-American treaty two years later extended trading rights at the five ports to the United States.[5] The opening of China to expanded trade and the continuation of illegal opium markets meant that transpacific excursions involving furs as a substitute for specie became unnecessary. Not all American merchants exchanged opium for Chinese goods, but by the mid-nineteenth century increasing numbers of them sought out China as a buyer of American industrial products. Textiles and other northern manufactured goods permeated the Chinese interior. By the 1850s the United States exported roughly $10 million worth of products to China and imported $30 million, although this represented only about 2.5 percent of all American foreign trade at the time.[6] Even though Asian markets never produced the commercial paradise many in the nineteenth century had envisioned, both opium and textiles helped assure the marginality of older trading patterns in the eastern Pacific and facilitated China's economic imbalance toward the West.

Encouraged by these events and global economic currents, sea otter traders bypassed the Far East as Canton proved an increasingly unproductive market. During the waning years of otter hunting in California, cargoes of furs were sent on hide and tallow ships to both Mexico and the eastern United States. For example, the American owners of the *Nymph* made $159 in 1841 from furs sent to Mexico. The *Admittance* purchased fifty-five pelts in San Diego in 1846 and sold them within one month after arriving in Boston.[7] Prior to the Alaska Purchase in 1867, rising numbers of Russian American Company furs were shipped to Europe instead of Kyakhta.[8] Sea otter pelts aboard American ships bound for the Atlantic ended up in the London fur market. Accounts of sales in the 1890s reported that otter pelts were highly valued, going for as much as £260. The *New York Times* reported that such high prices were paid "almost entirely [by] Russian noblemen, who especially prize the sea otter fur for the collars of their overcoats."[9] Hence while Russia lost access to much of its Pacific territory, American traders continued to supply customers with marine products from the region. Prior decades of China trading that carried furs west across the ocean and away from North America help explain why the United States never developed a domestic market for sea otter pelts in the nineteenth century, despite their oft-touted quality. According to one historian, most Boston Men themselves "probably never saw so much as the skin of one" sea otter.[10]

Hunting in the Northwest

One of the first to write about American sea otter hunting in the second half of the nineteenth century was Charles Scammon, who in 1874 wrote *The Marine Mammals of the North-western Coast of North America*. A sea captain who came to California in 1850 and worked in the whaling industry, Scammon enlisted in the United States Revenue Marine

Service (today's Coast Guard) during the Civil War and assisted with government studies of Russian America and Siberia prior to the Alaska Purchase. *Marine Mammals* represents the culmination of his work as an amateur naturalist. Scammon reported that American hunters out of California had begun pursuing otters in the Kurils and that the area between Grays Harbor and Point Grenville in Washington Territory constituted "the most noted grounds" north of San Francisco. Hunting in California, he claimed, "is no longer profitable for more than two or three hunters, and we believe of late some seasons have passed without any one legitimately engaging in the enterprise."[11]

Both Native Americans and whites took otters on the Washington coast. Local Natives used traditional spears and canoes, as well as rifles. Sold to white traders, pelts fetched hundreds of dollars apiece by the turn of the century.[12] Some hunters built wooden derricks or shooting stands from which they targeted sea otters. A runner on the beach recovered the carcasses, which sometimes had to be done before a waiting interloper snatched the kill. Victor Scheffer described one such location and the men who frequented it:

> In the 1870s, two men built a shooting stand on Copalis Rock; six years later Charlie McIntyre bolted the stand to the rock with iron bars. The remains could still be seen through binoculars in 1938. Samson John, with whom I spoke at Taholah, often took Charlie out to the rock by canoe and left him there for days at a time, where his only contact with the shore was by signal flag. Charlie later rigged a cable and pulley arrangement on which he could move a basket between the rock and the shore. He and his partner, Steve Grover, claimed to have killed 47 otters in one year. Charlie quit in 1903 when hunting ceased to be worthwhile.[13]

The combined methods of hunters proved all too effective, as sea otters were exterminated in Washington State and Oregon during the early twentieth century.

Local herds were concentrated along relatively limited stretches of coastline, which facilitated the building of derricks and hastened the destruction of the animals.[14] The last otter taken in Oregon was in the early 1900s, and the killing of several individuals in the early 1910s probably marks the extirpation of the species from the Washington coast (although exact dates for extinctions in the area are difficult to pinpoint in part because of regional migration).[15] The last pelts from the U.S. Northwest were probably sold in San Francisco and later found their way to the London market.

According to wildlife biologist Karl Kenyon, hunters in the early twentieth century also eliminated sea otters from the coasts of British Columbia and Baja California. In reference to a kill at Grassie Island, he concludes, "The sea otter apparently became extinct on the British Columbia coast during the 1920s."[16] The last sea otter in Mexico was reportedly taken in 1919, and Kenyon's visits there prior to the publication of his study produced no data on the animals, although otters have been sporadically reported along the Baja coast since the 1960s.[17]

The Alaska Purchase and the Alaska Commercial Company

The 1867 purchase of Alaska by the United States had a significant impact on sea otter populations in the late nineteenth century. Following improvements in U.S.-Russia relations during the American Civil War, politicians in St. Petersburg decided to deal away what remained of North American territories to the westward-expanding Americans. Perceptions of Russian America as an unprofitable fur territory and the turmoil of the Crimean War meant that the tsar and other royal officials were more concerned with securing the Amur River and holdings in the western Pacific. The idea of Alaskan settlements as crumbling and economically inefficient was perhaps more myth than reality, but it had influential proponents at the time.[18]

Though Secretary of State William Seward's offer of $7.2 million was ridiculed by some as "Seward's Icebox" and criticized as a political ploy to deflect attacks on President Andrew Johnson (whom the House of Representatives eventually impeached), a bill for the purchase of the territory finally passed Congress in July 1868. The same year saw the formation of the Alaska Commercial Company as one government-supported monopoly replaced another. The company took over fur industry operations in the territory and for a number of years assumed the responsibilities of civil government, similar to the Russian American Company. Alaska Commercial was essentially a merger of a number of competitors who had purchased trading stations and other properties and secured hunting rights from the Russian American Company prior to 1867. The mixed-race group of shareholders included San Francisco Jewish businessmen and a former employee of the Russian company.[19] By 1870 the Alaska Commercial Company had been awarded a twenty-year lease by the U.S. Department of the Treasury for the lucrative fur sealing operations of the Pribilof Islands, some three hundred miles off the western coast of Alaska in the Bering Sea. As part of the agreement, Aleuts there were guaranteed annual food rations, as well as school and health services. The Pribilof lease garnered huge dividends for investors during its run and brought at least $10 million in rental and pelt fees into the Treasury.[20]

While many Alaska Commercial Company records were destroyed in the 1906 San Francisco earthquake and fire, surviving evidence allows us to reconstruct its fur trade activities. In addition to the Pribilof enterprise, the company operated a number of trading stations throughout Alaska, although it had none in southeastern Alaska or on the Arctic coast. Native hunters were employed and offered items such as tea, coffee, clothing, and cookware to purchase from company stores. Western-style housing and education were provided, often with beneficial results for Alaskan people.[21] The company's largest trading and supply posts were at the islands Unalaska and Kodiak, and it was in those locations that the Alaska Commercial Company focused on sea otter harvesting. Decades of Russian American Company conservation practices had allowed otter populations to rebound from the destruction of the eighteenth century, and the Alaska Commercial Company willingly tapped these resources for existing markets (although according to some reports, Russian demand for otter fur was limited during the earliest years of the company's tenure in Alaska).[22]

The extent to which the Alaska Commercial Company exploited sea

FIG. 10. A photograph of a dead sea otter taken sometime in the late 1800s near Oyhut, Washington. University of Washington Libraries, Special Collections, UW39004.

otters for profit is somewhat more difficult to determine. Molly Lee writes that the increase of the Alaskan population "proved too gradual to allow the Company to focus solely on the sea otter harvest."[23] Moreover, the granting of the Pribilof lease had its genesis partly in concern about fur sealing competition in the late 1860s and its destructive outcome. Limiting access to one commercial outfit was seen as sound policy, and, as discussed below, it was pelagic sealing by independent hunters—not the Alaska Commercial Company—that posed the most direct threat to fur seals in the North Pacific by the late nineteenth century. Yet while a congressional attempt to restrict hunting activities in Alaska in 1868 listed the sea otter, the bill that granted the Pribilof lease two years later did not, despite its generic title, "An Act to Prevent the Extermination of Fur-Bearing Animals in Alaska."[24] As was the case with Russian marine mammal policy in the early nineteenth century, the abundant and lucrative fur seal herds that gathered in the Bering Sea took precedence over diminished otter numbers at the beginning of American sovereignty in the territory. This left *Enhydra lutris* largely at the mercy of indiscriminate hunting by both the Alaska Commercial Company and independent outfits, such as the one led by Kimberly onboard the *Cygnet*. Officials periodically acknowledged the delicate state of Alaska's sea otters after the 1868 statute, but enforcement lagged and ultimately proved too difficult for local authorities.[25]

Alaska Hunting

Hence it was both Alaska Commercial Company vessels and independent ships that carried hunters and their *baidarkas* to grounds in the Aleutian Islands and elsewhere in the territory in search of otter herds. A report from the 1890s on the local industry identified the company as responsible for facilitating the majority of hunting, but others played a role.[26] Traditional killing methods were increasingly supplemented with firearms by the 1880s, resulting in "surf-shooting" of animals from shorelines (similar to hunting in Washington and Oregon). Before that time, Native Alaskans had been denied the right to own breech-loading rifles by government decree.[27] The combined techniques proved too much for sea otters to bear. One contemporary record lists thousands of sea otter skins collected per year from the Aleutians and Kodiak, peaking in 1885 at just over four thousand. Ten years later only 887 skins were reported.[28] One of the few vocal advocates for sea otters in the late nineteenth century was the artist and self-made naturalist Henry Wood Elliott. In his work he contrasted the high 1880s totals with lower catches during the waning years of Russian sovereignty. Linking the environmental damage of his day to what occurred over the course of the eighteenth century, Elliott concluded that "the difference is not due to the fact of there being a proportionate increase of sea-otters, but . . . to the organization of hunting parties fired by the same spirit and competitive ardor as that which animated and shaped the hunting during early days of Alaskan discovery."[29]

Extensive schooner- and land-based hunting of sea otters in Alaska by the last quarter of the nineteenth century was facilitated by authorities lifting a Treasury Department regulation from 1879 that attempted to limit hunting in the territory to Natives or whites married to Native women. The restriction proved ineffective because a sizable number of white immigrants in Alaska married to avoid the restriction. Rescinding it encouraged competition and reportedly prompted the Alaska Commercial Company to grudgingly expand its own sea otter operation.[30] Similar to their Russian predecessors, American enterprises relied heavily on Aleut skill to extract marine mammals from nearshore areas, and they stocked trading stations to exchange skins for goods desired by hunters. Additionally, crews on schooners like the *Cygnet* often located otter rafts opportunistically and collected both seals and otters when they could. Kimberly's noteworthy vessel notwithstanding, information on such vessels is difficult to uncover because sea otter hunters were often taking fur seals illegally and thus were encouraged to enter Alaskan waters without leaving much in the way of written records behind.[31]

As various maritime and stationary hunters preyed upon North Pacific wildlife, indigenous inhabitants of the former Russian territories were left without a vital source of income. C. L. Hooper noted in an 1897 Treasury report that a number of the Alaska Commercial Company's stores were shut down due to declining catches. Individuals in the western Aleutians were being transported by schooner to more productive hunting grounds to the east, particularly around Kodiak Island. Hooper warned that the sea otter needed protection not only for its own sake but "for the benefit of the natives who have been dependent upon it for more than a century, and who will be reduced to suffering and want without it."[32] Thus communities like Belkofski in the eastern Aleutians that were for a time enriched by the American hunt experienced economic depression by the turn of the century.[33]

Pacific people also continued their traditional pursuit of sea otters along the coasts of southeastern Alaska and coastal British Columbia in the late nineteenth century. As the Russian American Company historian P. A. Tikhmenev wrote in the early 1860s, "At present only the natives of Yakutat and L'tua [Lituya] Bays hunt sea otters, because there are more otters there than in other places along the northwest coast of America."[34] United States Navy lieutenant George Thornton Emmons, who conducted ethnographic work in the area at the turn of the twentieth century, noted that four otters were killed at Cross Sound in 1889 and that Hoonah people took some sixty animals at Lituya Bay in 1892.[35] Evidence of hunting along these coastlines appears less frequently than for other locations, in part reflecting the historic damage that had already been done to sea otters in the Pacific Northwest. Nevertheless, low-level hunting gradually reduced remaining populations into the early twentieth century. Unlike in California, which also experienced a gradual decline, otters in southeastern Alaska and western Canada faced Native American groups who traditionally killed them, an important factor in the extirpation of the species there during the first few decades of the 1900s.

The last Alaskan schooner hunts were curtailed by regulations in 1906 that required vessels to register as foreign ships for customs purposes and banned hunting within nine miles from shore. Amchitka Island and the Kodiak area continued to support some outfits, but such excursions became increasingly rare. As Cal Lensink notes, "In 1910, a crew of 40 Aleut hunters on the two American vessels left in the sea otter trade obtained 7 otters, and 11 others were killed or found dead on beaches for a season's total of only 34 animals."[36] The following years saw the legal end of the indiscriminate killing of sea otters by Americans.

Hunting in the Kuril Islands

As Pacific whaling declined in the 1860s, some British and American mariners plied the Kurils seeking sea otters as commercial substitutes. This activity increased dramatically after the *Cygnet*'s visit to the islands in 1872. Schooners soon swarmed the area, benefiting in part from the political instability caused by Russo-Japanese tensions that preceded the Treaty of St. Petersburg in 1875. While the treaty placed the archipelago under Japanese sovereignty, prior concessions of extraterritoriality further assisted foreigners in their pursuit of marine mammals in the western Pacific. Diplomatic conditions and the determination of hunters contributed to Japan's inability to control what was effectively foreign poaching within national waters.³⁷

H. J. Snow, an Englishman working at a firm in Yokohama who heard of Kimberly's fur take, decided to take up hunting in the islands. He became the most well-known otter hunter of the Kurils, writing two books about his exploits before his death in 1915. Snow and his crews launched small boats from larger vessels and killed the animals utilizing a triangulation method similar to the one used by Aleut hunters in *baidarkas*. Most men who served with him were Japanese, although at one time he employed a number of Chinese sailors. Kuril hunts of this era were time-consuming and often very dangerous affairs. Perilous sailing conditions and misaimed rifle shots accounted for a number of injuries and deaths.³⁸ Neither Japanese nor Russian officials—the latter concerned with sea otter and fur seal poaching in the Commander Islands—were able to completely stop these hunts, despite their illegality, even though Snow himself was arrested once by officials from both nations. Additionally, some Japanese and Russian schooners also harvested otters. Accordingly, herds were depleted throughout the archipelago by the 1890s. Snow observed, "The old schooner Diana was the last to quit the business. She kept at it until 1893, when she also gave it up as being no longer a paying venture."³⁹

The Kurils provided as much as one-fifth of the American otter fur take by the early 1880s. An 1881 article in the *Pacific States Watchman* noted, "The number of sea otter skins taken annually is not definitely known, but from the most authentic information we can obtain, the aggregate for the past three years has been 5,000, 1,000 of which came from the Kuril islands; and, valuing each skin at $50, amounts to the sum of $250,000."⁴⁰ Sea otters in both the Kurils and the Aleutians survived the onslaught of the late nineteenth-century hunts in part because of the geographic and oceanographic contexts of the island chains. Both are relatively isolated and difficult for even modern sailors to navigate. Fog was particularly troublesome for Kuril hunters. Snow noted at one point that out of an eighty-two-day stint in the islands, sixty-five days had been too foggy to find sea otters.⁴¹ In the early 1900s a sizable remnant population existed at the Sanak Islands (just south of the western edge of the Alaska Peninsula), and the area's especially dangerous reefs and rocks played a part in the otters' survival.[42] Even as laws protecting marine mammals were beginning to strengthen by the turn of the twentieth century, the foreboding nature of Pacific habitats provided nearshore spaces where some animals could escape death. Biologist Linda M. Nichol writes that in Alaskan locations "law enforcement may not have been as big a factor as the great distance from ports, severity of the climate, and the scarcity of the animals."[43]

At the Commander Islands, sea otters were still being managed by Russian harvesting measures while foreign schooners attempted to exploit the local population. At Medny Island (also called Copper Island) in the early 1900s, one to two hundred animals were reportedly taken each year by Native hunters under strict management guidelines, which included not using guns or building fires near the shore and releasing any females or young caught in nets.[44] But overhunting and poaching led to declines by the 1910s. By 1919 the local catch was reduced to thirty-five, and five years later the Soviet government banned otter harvesting at the Commanders.[45] Despite the crash, the near century-long Russian head start in marine mammal regulation enabled individuals to make important advances in conservation biology by studying otters in captivity at the Medny and elsewhere, as noted below. The proverbial birthplace of the fur rush in the North Pacific produced beneficial scientific knowledge for the species in the early twentieth century.

Hunting in the Californias

Like their cousins in the Kurils and Aleutians, some California otters benefited from living in relatively remote places even as their overall numbers gradually declined. George Nidever, who hunted in Mexican California beginning in 1834, continued to take sea otters until 1855.[46] Two of his sons, with Santa Barbara businessman Eugene Rogers, plied the coast along Baja and the Channel Islands with schooners in the 1870s and early 1880s. Elliott cited a report on the Rogers outfit, which was also engaged in abalone and other marine mammal harvesting, claiming that it took seventy-five otter skins in the Channel Islands in 1880.[47] While organized pursuit continued throughout the region for a number of years—with Baja hunting engaged in largely as an adjunct to supplying the booming mining communities there after 1850—the California population was generally spared the type of widespread predation seen in other areas of the Pacific.[48] In part this was because the animals had been greatly reduced prior to the mid-nineteenth century and the arrival of miners. The gold rush likewise played a role in shifting the priorities of would-be hunters. The few men left on the coast by the late 1840s were at least for a time drawn to the diggings and away from the fur trade. In 1848 the senior Nidever and his crew heard the news of gold. As he later recalled, "The prospect of getting $16 a day when their monthly wages barely amounted to that was too great a temptation for our men, who insisted on leaving us at San Francisco."[49] Nidever joined them on the trip up the Sacramento River and to the camps.

More importantly, the pursuit of sea otters elsewhere in the Pacific diverted attention away from California. Both Scammon and Snow noted the prevalence of California-based ships in the Kurils in the 1870s. Snow recalled that in his first season in 1873, there were six other schooners in the business, all of which hailed from California. The following year near Iturup he counted nine from the state and three from Japan.[50] Like the *Cygnet* before them, many of these vessels carried independent merchants seeking opportunities for profit in various marine products. Snow described Kimberly as "an old otter-hunter" who originally sought to salvage whalebone from ships frozen in the Bering Strait and was planning to search for fur seals near Hokkaido.[51] As with the Kurils, many if not most of the independent schooners that came to Alaska were California vessels, although both maritime and stationary hunters and traders traveled to the territory from throughout the eastern Pacific.[52]

Thus as individuals turned their attention to the north and west of California, surviving otters in the Golden State were afforded a measure of respite. To be sure, as demonstrated by maritime data recently compiled by Marla Daily of the Santa Cruz Island Foundation, a series of hunting vessels still plied the central and southern California coast in the late nineteenth century, particularly at the Channel Islands, and sea otters were exterminated from Baja by the early twentieth century. The records contradict the notion that the animals were absent from the Channel Islands after 1850 and suggest that San Miguel Island was home for a time to a relatively substantial remnant population.[53] Chinese fishermen in the region contributed to the late hunting efforts. In 1885 a Santa Barbara newspaper reported a Chinese junk at anchor, writing that it was "presumed by the Captain that they will be instructed to go south in sea[r]ch of otter and whale."[54] Still, such activities were distinct from the more focused hunts seen in the Aleutians and Kurils. Had California witnessed a similar rush of vessels, then sea otters might have been eliminated along the entire coastline south of the Strait of Juan de Fuca in the early twentieth century. Instead, the half century of hunting prior to U.S. sovereignty made such pursuit commercially unfeasible. California otters also benefited from being dispersed over a wider geography than in Washington State, which effectively discouraged stationary hunting and helped preserve remnant populations.

Fears of Extinction

By the beginning of the twentieth century many feared that *Enhydra lutris* was on the verge of total extinction. An 1899 report on London fur sales informed buyers that "with a probability of a greatly reduced supply next year, and the possibility of the animal's early extinction, the advance is 50 per cent."[55] The famed Harriman Alaska expedition also expressed a negative assessment for both popular and scientific audiences. A Gilded Age railroad executive from New York (and later president of the Union Pacific Railroad), E. H. Harriman sponsored an impressive fact-finding venture to Alaska to accompany his family on a western vacation. When the expedition left Seattle in May 1899 it boasted more than two dozen scientists, artists, and photographers, including the California-based naturalist John Muir.[56] The specialists onboard Harriman's steamship realized that they were engaged for the most part in reconnaissance and not in-depth study of the territory, although the volumes they wrote in subsequent years made distinct contributions to science. Grove Karl Gilbert's *Glaciers and Glaciation* was perhaps the most important volume in the expedition's series and significantly advanced knowledge of glacial geology.[57]

The Harriman reports said little about North Pacific marine mammals. While this is partly due to the fact that the expedition's coordinator and mammal expert, C. Hart Merriam, was kept busy encouraging other scientists to compile their information, the lack of animals on which to report was also a reason. As expedition geographer Henry Gannett surmised, "The natural resources of Alaska are enormous. . . . Some of these, however, have begun to suffer from the drain to which they have been subjected. The gathering of furs and skins, which has been in progress since the early Russian occupancy[,] . . . has been prosecuted so actively that the fur trade is now of comparatively little consequence. . . . The sea otter has become very rare, and the fur seals . . . are now reduced to a small fraction of their former number."[58] According to historian Kathryn J. Frost, the expedition rarely encountered sea otters.[59] Before the voyage ended Harriman himself managed to buy a pelt from a local seller. Otherwise the details of his Alaska adventure left readers with the impression that otters were close to disappearing. The Alaska Commercial Company itself concurred. One company official observed, "The beautiful sea otter . . . is practically extinct."[60]

The North Pacific Fur Seal Convention

Progressive Era conservation came to the aid of otters and seals. At the turn of the twentieth century, a broad movement of preservationists and utilitarian reformers worked to support a variety of natural resource causes in the United States. Advocates often had conflicting motives and goals yet succeeded in enacting state, federal, and international statutes meant to protect disappearing wildlife. For Pacific marine mammals, the most notable support came with the North Pacific Fur Seal Convention of 1911. Unlike the Alaska Commercial Company's Pribilof management lease some forty years prior, the 1911 treaty specifically included the sea otter, a landmark recognition of the historic connections between the species. Hunters often turned to seals to supplement declining catches of otters in the nineteenth century; now Progressives included otters in their attempt to save the seal. Unfortunately, the protracted campaign for fur seal conservation distracted from the sea otter's dilemma at least as much as it helped to highlight it. As seen above, 1911 came too late to save some local populations. The widespread killing of the animals would eventually end, but the priorities of turn-of-the-century officials were too often elsewhere.

The dramatic decline of the fur seal in the Pacific received more attention in part because it occurred during this era of national and international environmental awareness. Additionally, unlike otter pelts, seal skins found markets both within the United States and abroad, which energized American interest in ensuring the sustainability of the industry. The crisis began during the 1880s. Despite the management efforts of the Alaska Commercial Company the herd at the Pribilofs showed signs of weakening. Pelagic sealing—killing the animals at sea, including during their travels from wintering grounds off California—was an extremely destructive practice, yet it was on the rise, particularly with commercial groups in Victoria and Seattle. Sealing schooners lost 50 to 80 percent of the dead creatures from sinking after being shot. Mother seals and young females were especially vulnerable because they spent more time in the summer at sea, the former leaving island rookeries for international waters to find food for pups. Hence pelagic hunting decimated the breeding population and left future generations starving to death in the Pribilofs.[61] When in 1886 U.S. Revenue Marine Service ships began capturing Canadian sealing vessels in international waters, a diplomatic crisis was born. The marine resources of the Pacific once again spurred geopolitical conflict. Sealers contested national boundaries in their slaughter of quarry that ignored such artificial markers, and the United States ironically found itself in a position in Alaska not unlike what Russian officials encountered with American traders in the early nineteenth century. This time, conserving nature became central to an international resolution.

At a Paris tribunal in 1893 American diplomats initially focused on a traditional argument involving seals as national property. The United States lost this case and was ordered to pay damages for taking Canadian schooners, but a sixty-mile buffer zone around the Pribilofs was ordered to be free of sealers, and a ban on the use of firearms in pelagic sealing was agreed to.[62] Still, without a complete ban on the pelagic hunt the Bering Sea herd continued to decline. Moreover, Japanese hunting outfits (subsidized by their government) began to take a larger role in the controversy. Animals that numbered over 2 million in 1870 stood at some 140,000 as the century drew to a close. In response, the United States banned pelagic sealing by Americans, called for talks involving Japan and Russia, and began to rely more heavily on scientific arguments to convince foreign officials to end the practice. Chief among those working to build a case against pelagic sealing but for continued land sealing was David Starr Jordan, president of Stanford University. Jordan was chosen by the Treasury Department to edit a four-volume study titled *Seal and Salmon Fisheries and General Resources of Alaska*, published in 1898. His utilitarian approach contrasted with the preservationist and sentimental impulses of the marine mammal advocate Elliott, and public clashes between the two in the years that followed, fueled in part by Elliott's sensitivity to criticism and his need for official recognition of his natural history work, may have delayed resolution of the issue.[63]

Although scientists and politicians included the sea otter in their deliberations, the emphasis was on fur seal conservation, and many felt that otters mattered only insofar as their tragic state foreshadowed what might happen to the seal. Jordan's collection of data on Pacific marauders like the *Cygnet* and the Treasury documents that he incorporated left the impression that such vessels were significant not because they were otter hunters but because they killed seals. Thus whatever legal structures might have been able to protect the animals around this time were effectively ignored. Treasury agent Joseph Murray was candid about the weakness of American support for the species: "Thus for a quarter of a century did the United States throw every possible safeguard of law around the seals and other fur animals of Alaska . . . while during the same period of time the sea otter, which, owing to its pelagic habitats, was necessarily left to the tender mercies of the pelagic hunter, who knows no law higher or holier than avarice and selfishness, has been practically exterminated."⁶⁴

The plight of Pribilof Aleuts in the wake of pelagic sealing was also a focus of concern. Pegged to harvested skins, Aleut real wages fell after 1890 under the new government lessee, the North American Commercial Company, which took over from the Alaska Commercial Company. Declining fur seal numbers coupled with sustained high prices for company goods on the islands resulted in government subsidies to supplement Native income beginning in 1892.⁶⁵ Some officials complained of this expanded role in the welfare of hunters, since goods were subsequently allotted weekly with no say from the purchaser. One Treasury agent wrote that the new payment method "prevents the progress that accrues from the superior skill or greater self-denial and makes a virtual almshouse of the Pribilof reservation."⁶⁶

Following a preliminary Anglo-American ban on pelagic sealing in 1911, both Russia and Japan sent diplomats and scientific advisors to a four-power conference in Washington, DC. After weeks of negotiations and the prodding of the Japanese emperor by President William Howard Taft, the American proposal that took years to develop was officially accepted. Canada and Japan each received $200,000 and 15 percent of takes from American and Russian land sealing, and all parties signed a ban on hunting at sea. According to Kurkpatrick Dorsey, the North Pacific Fur Seal Convention succeeded "because each of the four countries accepted, sometimes grudgingly, the scientific and moral evidence that pelagic sealing had to end, even though it was legal practice."⁶⁷ Reformers, politicians, and determined diplomats deserve a great deal of credit for this and for the subsequent rebounding of fur seal numbers. The sea otter was included in Article V of the treaty: "Each of the High Contracting Parties agrees that it will not permit its citizens or subjects or their vessels to kill, capture or pursue beyond the distance of three miles from the shorn [*sic*] line of its territories sea otters in any part of the waters mentioned in Article I of this Convention."⁶⁸ Yet by banning hunting only within international waters governed by the terms of the treaty and not in the nearshore habitats where the animals lived, the

Fur Seal Convention effectively did very little for otter conservation. As VanBlaricom notes, "One might reasonably argue that the Treaty was *never intended* to provide conservation value in an ecological context."[69]

Otters were not simply victims of fur seal diplomacy and conservation. But greater effort prior to 1911 might have reduced the damage done by late nineteenth-century depredations. Given that most hunting was by American outfits and that after 1867 much of the sea otter's habitat was within the territorial waters of the United States, conservationists could have underscored the long history of killing otters and the need for limiting the practice independent of international entanglements regarding seals. This is essentially what Hooper proposed in his Treasury report of 1897, but by then it was likely too late to avoid localized extinctions.[70] Unfortunately, the species in many ways became a victim of its own bloody past. Individuals like Elliott romanticized the sea otter's natural history and its historic plight. In his works he marveled at the accomplishments of Vitus Bering and promyshlenniki and colorfully described Aleut hunters while lamenting their treatment under Russian masters. Such dramatic framing had the effect of presenting the animal as a lost artifact from a bygone era, softening any message of environmental urgency. To his credit, Elliott was conscious of this literary dilemma: "A truthful account of the strange, vigilant life of the sea-otter and of the hardships and perils encountered by its human hunters would surpass in novelty and interest the most attractive work of fiction. I mention this with much emphasis, because throughout the following narrative many instances will arise, coupled with the life and chase of the sea-otter, which may strike the reader's mind as the evolution of romantic thought."[71]

Seen in this light and coupled with pronouncements concerning the imminent demise of the species, Article V of the Fur Seal Convention can be read as both a united call for saving the sea otter and a short obituary for part of the Pacific's biological past. Despite conservation inadequacies, however, the sea otter continued to receive important legal help in the early twentieth century. The Aleutian Islands Refuge was established in 1913, offering federal protection of the primary habitat of Alaska's remaining population. That same year California banned the killing of otters. Some restrictions on hunting were put in place by the government of Japan in 1912. Otter hunting was not officially prohibited in Canada until 1931.[72] Altogether such protective measures provided necessary safeguards that allowed the animals to rebound in areas of the North Pacific. In the United States the Alaska population recovered most quickly. Naturalist Olaus Murie, working for the U.S. Bureau of Biological Survey, noted seeing otters on the Alaskan Peninsula as early as 1925, and his expeditions to the Aleutians in 1936 and 1937, according to Kenyon, "were the first to reveal that otter populations at a number of islands were growing."[73] By midcentury Alaskan sea otters numbered around ten thousand individuals.

Rediscovery

Sea otters recovered at the Commander Islands to the point that pioneering attempts were made by Russian researchers in the 1930s to keep the animals in captivity for both scientific knowledge and the purposes of the fur industry. At Medny Island in 1932 an abandoned building with a freshwater pool was used to house Fomka and Bek, two sea otters kept separately and observed for nearly two weeks, during which time both died from gastrointestinal inflammation.[74] Subsequent efforts were made to study captive specimens at Medny, and in 1937 Soviet researchers attempted the first—and in many ways the most dramatic—sea otter translocation, moving animals captured from Medny by ship and railroad to the Murman shore of the Barents Sea in northwestern Russia. Nine individuals were initially shipped, but only two made it to Vladivostok alive. In November 1937 the remaining two otters arrived by rail cart at Yarnyshnaya Bay, where they lived together in two different seawater-infused enclosures until January 1940, when one of the animals, named Buyan, escaped and reportedly survived in the bay for more than two years. Yashka, the other translocated otter (both were males), was euthanized after Buyan's escape due to the Soviet need to focus on World War II.[75] Despite the cancellation of the program, marine mammal science was advanced in Stalin's Russia just prior to the "rediscovery" of the sea otter on the western coast of the United States.

In the early twentieth century low-level hunting of sea otters in California restricted them largely to the central-state coastline of Monterey County. In 1915 an article in *California Fish and Game* reported on sightings near Monterey and Point Sur. Exaggerating a bit yet expressing confidence in the recent environmental statutes, the author wrote that "enforcement of [the 1913 state] law appears to be greatly benefiting the species."[76] Two more reports were included in the same journal in 1917 describing two sea otters sighted in the Monterey area and thirty-one animals south of Catalina Island.[77] Even if the Catalina group was later extirpated,[78] by 1920 a handful of authorities understood that California sea otters were recovering. Partly because such reports—as well as Murie's studies in Alaska and the Russian captivity experiments—were not widely known prior to 1938, news of a sizeable population along the Big Sur coast was susceptible to mischaracterization.

In March of that year Howard Granville Sharpe saw sea otters at the mouth of Bixby Creek in Big Sur. In a colorful account written in 1939, Sharpe described how he and others peered through a telescope from the porch of his seaside ranch and were astonished at what they saw. He equated sighting hundreds of otters to discovering the extinct dodo and finding dinosaur eggs in Mongolia.[79] The genuine surprise of some officials who were called to investigate was due more to the size of the herd at that time than to the mere fact that the animals were present. A report in the *Monterey Peninsula Herald* for March 28 confirmed that Fish and Game wardens had been aware of the otters' presence for years.[80] Sharpe's desire for self-promotion and monetary gain apparently motivated his revelations. The completion of a portion of Highway 1 in 1937 connecting the area to San Luis Obispo to the south promised to bring more visitors to the remote location and across the recently constructed Bixby Creek Bridge—precisely where Sharpe and his wife owned land and a small tourist stop.[81] Yet even if he wished to profit from breaking the local silence regarding the otters, the publicity that he helped generate brought attention to preserving them.

The Monterey area happenings were also dramatized in a pamphlet written at the time as a "Phoenix-like appearance" of the sea otter that "astounded everybody."[82] Such hyperbolic responses worked to magnify the animal's profile and cemented 1938 as a landmark year in marine mammal history. Over time, additional mythmaking regarding the events has worked to diminish the size of the local herd—some popular sources today cite *fifty* animals as having been present, an error that may have evolved from an original Monterey newspaper article that identified "several hundred of them in herds of about 50 each."[83] The image of a small group of creatures clinging to life on a remote California coast makes for an uplifting tale of wildlife recovery, but it obscures some of the early twentieth-century environmental successes that took place prior to 1938 and needs correction. By the 1950s the California population numbered approximately five hundred individuals, and rafts of young male vanguards had begun to work their way north toward Monterey. In 1962 they were recolonizing the bay.[84]

Despite these positive developments, *Enhydra lutris* recovered more sluggishly in California than it did in Alaska in the first half of the twentieth century. Biologists, ecologists and other researchers continue to analyze the complex factors involved in sea otter decline and recovery, so while definitive explanations for slow population growth are difficult to ascertain, a few general points can be made. If James Estes, Brian Hatfield, Katherine Ralls, and Jack Ames are correct that otter mortality "appears to be the main reason for both sluggish growth and periods of decline" for the California population, then how did the local animals die in the early 1900s?[85] Continuing human predation following the Progressive Era statutes is one likely answer. Once again, geography must be considered, since it was simply easier for hunters—and then poachers after 1913—to shoot otters along California beaches than it was to track them down in the Aleutians. Kenyon claimed that poaching remained a larger concern in California than in Alaska at midcentury.[86] It is reasonable to assume that the same held true earlier and slowed the animal's growth in the state. The U.S. government held numerous auctions for confiscated skins throughout the 1920s and 1930s, but determining how many of these came from California versus elsewhere needs more attention.[87]

Humans are not the only otter predators in California. Bitten carcasses are found periodically on Golden State beaches, confirming that great white sharks attack them and play a role in local recovery. The number of identifiable shark kills in California has varied year to year in recent decades but is significantly higher during times of otter population decline.[88] Still, it is difficult to know the extent to which great whites might have been a limiting factor in the early twentieth century—the Bixby Creek herd was reportedly assaulted by a killer whale,[89] a marine creature that proved to be an effective killer of Alaskan sea otters at the turn of the millennium (discussed in the following chapter). Disease has been identified as another key factor in sea otter mortality, and the dense human population of California may produce higher rates of disease-related death. Estes and colleagues note that a variety of infectious diseases are "the proximate cause of death in more than 40%" of fresh otter carcasses recovered in California, and research suggests that "organisms associated with humans and domesticated animals are significant contributors to disease-related mortality in sea otters."[90] Though changes in this percentage apparently are not associated with recent periods of growth and decline, the authors conclude that "infectious disease may well be an important factor for the overall depressed rate of population growth in California sea otters during the 20th century."[91]

The dynamic Pacific ecologies to which *Enhydra* is linked remind us to be cautious when examining historical data alongside biological and oceanographic trends. Despite challenges, such an approach will be important for shedding future light on a crucial era in marine mammal history. Preyed upon for decades in the late nineteenth century, fur seals and sea otters began to receive necessary legal benefits in the early 1900s. Protections made environmental recovery possible in areas of the North Pacific. For sea otters, the realization that the killing needed to stop, no matter how overshadowed it was by other economic and political interests, ultimately proved effective. Countering regional and global developments that had once threatened to doom the species, national and international political efforts helped saved it. Competition gave way to cooperation at the beginning of the twentieth century, sparing the sea otter the fate of the sea cow.

FIG. 11. California Fish and Game wardens examining a sea otter pup in February 1950. J. B. Phillips Collection, California History Room, Monterey Public Library.

5   Nukes, Aquaria, and Cuteness

Cynthia Holmes was a student at the University of British Columbia in 2002 when she visited the Vancouver Aquarium and shot a one-minute-and-forty-one-second video. She kept the digital images for five years until she decided to upload them to a new Internet site called YouTube in March 2007. The video that Holmes shared became an international sensational within weeks, making stars out of two sea otters at the aquarium named Nyac—a survivor of the *Exxon Valdez* oil spill of 1989—and Milo. The animals are "holding hands" at the beginning of the clip (emphasized by a quick zoom on their clasped paws), which prompts responses such as "Absolutely adorable!" and "Cute!" from observers at the aquarium. At the fifty-five-second mark the otters float apart, only to drift back toward each other and reconnect their paws some twenty-five seconds later, followed by a chorus of "aww." One comment from a YouTube viewer posted on April 8, 2009, cynically captured the video's gushy sentiment: "My head will explode from cuteness." *Otters Holding Hands* was viewed 1.5 million times in just two weeks after Holmes uploaded it, making it YouTube's most popular animal video at the time. According to a report on the phenomenon, "Aquarium officials said they hope the internet audience learns that the otters are not only cute, but are an endangered species as well." Images and videos of sea otters holding hands have outlived Nyac and Milo as social media memes. The otters had a book of poetry dedicated to them by a Canadian author in 2013.[1]

The encounter between Nyac, Milo, and Holmes at the Vancouver Aquarium was made possible by a number of crucial developments in sea otter history since the middle of the twentieth century. The establishment of otter exhibits at zoos and aquaria and environmental tourism are part of this broader story, as are man-made ecological disasters like the one that occurred at Prince William Sound, Alaska. These dynamics in turn intimate the role played by national and international politics involving the Pacific Ocean at the turn of the millennium. The threat of an oil spill to sea otters made for dramatic headlines and stirred debate in 1989, but so did nuclear explosions in Alaska some two decades prior. While not often appreciated as a key factor in sea otter conservation, the effects of U.S. atomic tests at Amchitka Island in the 1960s and 1970s figuratively radiated across the ocean, as the animals were translocated to coastal habitat from which they had been eliminated during the maritime fur trade. All of these historical currents flowed toward Vancouver in 2002 and helped result in a viral video that spanned the globe. Yet Nyac, Milo, Holmes, and their YouTube viewers share in a cultural experience that is more recognizable but just as complex: sea otter cuteness. The natural charisma of the species as mediated through a variety of contemporary media outlets and local tourism industries has helped to bring wider public attention to *Enhydra lutris*, even as commodifying it in nondeadly ways troubles some conservationists.

The recovery of the sea otter since the twentieth century is often written about as a praiseworthy environmental triumph even as its delicate present status is emphasized. The goal here is not to summarize current literature on the state of the animals. Instead, this chapter analyzes selected events, individuals, and controversies in contemporary otter history. To an even greater extent than the record of late nineteenth-century near extinction and early twentieth-century conservation, scholars have been largely silent when it comes to narrating the sea otter's recent past. Though the human and animal stories included in this chapter represent but a few relevant topics, their historical and biological relevance is hopefully evident. For example, as noted below, recent decades have demonstrated that, whatever the causes, dramatic otter decline is not an artifact of bygone eras. Additionally, the tangled cultural web in which the species finds itself today needs scrutiny. Why have the marine creatures, virtually forgotten at the turn of the twentieth century, become so captivating to so many people? Biologist Glenn VanBlaricom suggests that we may never know the answer when he writes, "There is a short list of mammals whose inexplicable appeal overwhelms our best efforts to be logical and dispassionate.... Sea otters are surely on the list."[2] As inexplicable as cuteness and its by-products may appear, historical and cultural answers are pursued here. Input from and collaboration between conservationists, sociologists, psychologists, and others can guide further analysis. As this book has argued, the iconic animals deserve broad, interdisciplinary attention. Naturalists and fur trade historians should be joined by others committed to dispassionate studies of them, for how we think about sea otters is as important as thinking about them more often.

Amchitka's Atomic Diaspora

World War II disrupted marine mammal conservation efforts in the Aleutian Islands. Air strikes, shellings, and construction projects following the Japanese invasion of the archipelago created considerable damage to coastal habitat. Additionally, many American servicemen engaged in recreational hunting during the war. One navy admiral complained that skins were showing up as presents to "relatives [of personnel] in Seattle . . . in violation of existing law."[3] After the conflict, refuge manager Robert "Sea Otter" Jones became an important advocate for conservation at the islands, gaining his colorful nickname from his indefatigable spirit and the work he conducted at Amchitka Island, which was home to one of the largest local populations of sea otters by the 1950s.[4] Jones had served in the war as an army radio and radar operator at Adak and Amchitka, helping to monitor Japanese aircraft

FIG. 12. Robert "Sea Otter" Jones arriving in Seattle with a sea otter from Amchitka Island in June 1954. CriticalPast.

and naval forces. He witnessed disastrous environmental effects in the Aleutians and afterward decided to stay in Alaska to conduct biological work. The U.S. Fish and Wildlife Service officially hired Jones in 1948 to manage Aleutian wildlife.[5]

Jones was motivated to fight against U.S. plans to test nuclear weapons at Amchitka. It was deemed a suitable location for such Cold War efforts despite a partial victory by Interior Department officials to limit military encroachment in the refuge following the war. Jones's complaint to his Interior Department superiors in 1950 helped lead to the cancellation of testing plans at the island the following year.[6] Meanwhile, North American natural history writing about the sea otter began to focus more on its charismatic and human-life qualities, helping to bring increased public attention to the species. While Jones wrote in a 1951 report that the animal "endears itself to the average human being because of its habits," he and other researchers at the time tended to avoid excessive romanticism, leaving such prose to popular authors who had increasingly large postwar audiences.[7] For example, a 1949 article in the *Saturday Evening Post* about Alaskan otters provided details on the "playful," "gregarious" creatures and included dramatic descriptions of mothers and pups. The sentiments were included alongside predictions of the return of hunting the Alaska population for the fur trade.[8] Natural history media likewise found it difficult to resist highlighting the appeal of the species. A 1958 article in *Outdoor California*—a magazine produced by the California Department of Fish and Game—titled "The Playful Sea Otter" included a photograph of a taxidermied sea otter holding a shellfish, underscored with the caption, "Even a mounted specimen looks playful."[9] It was an early example of an otter photograph joined with an anthropomorphic description. The *Saturday Evening Post* article's image of live Alaskan representatives was captioned with economic data: "The return on the investment for catching them once ran to 600%."[10]

Despite initial success at stopping nuclear experiments at Amchitka, the threat of underground testing at the island had returned by the middle of the 1960s, due in part to the government's need of being able to distinguish seismic activity from atomic explosions as required by the Nuclear Test Ban Treaty of 1963. Protests against the new plans reached international proportions and included outcries from Canadian politicians and activists. By the end of the decade, the Don't Make a Wave Committee had formed in Vancouver, its name inspired by fears regarding potential Pacific tsunamis that could be set off by underground detonations in the North Pacific. In 1972 the organization officially became known as the Greenpeace Foundation, having previously launched an unsuccessful maritime mission to force a halt of the final Amchitka test.[11] According to D. J. Kinney, the various Amchitka controversies of the 1960s and early 1970s helped instill within the modern environmental movement a transnational social and political energy that united a broad coalition of groups around a central goal. The responses to the crises involving the island, he writes, "served as a blueprint for the widespread transnational activism that would follow in the 1970s and 1980s."[12] Yet if concern over government secrecy and nuclear policy animated the protests and shaped environmentalism more broadly, ecological fears regarding nukes remained central, and potential and real damage to Amchitka's sea otters was often the focus of attention at the time.

For example, in October 1965 Project Long Shot became the first atomic detonation at the island and occurred despite fears expressed in *Life Magazine* that it would "do irreparable damage to sea otters."[13] Shot Milrow in 1969 was followed by Shot Cannikin in 1971. Prior to Milrow, the national Sierra Club's member newsletter reported on anxieties regarding both seismic activity and otters, noting, "Alaskans are concerned that the blast will trigger earthquakes, and conservationists fear irreparable damage to the 56-year-old sea otter and water-fowl refuge."[14] While apparently few if any otters died from the first two tests, Cannikin—the largest ever underground nuclear experiment by the United States—killed anywhere from several hundred to as many as one thousand sea otters from rockfall and shockwave.[15] In response to mounting pressures, the U.S. Atomic Energy Commission funded successful translocations of sea otters in the late 1960s. Earlier attempts in the 1950s (led in part by Jones) to relocate otters from Amchitka to elsewhere in the Aleutians and to the Pribilof Islands had failed due largely to poor husbandry methods.[16] Beginning in 1965, the first Amchitka translocation representatives survived beyond their home at St. George Island in the Pribilofs and in southeastern Alaska. Groups were later moved to Prince William Sound, British Columbia, Washington, and Oregon. All but the Oregon translocations are considered to have been successful despite the fact that a majority of all transported sea otters died sometime after being released.[17] According to population studies, approximately 35 percent of all sea otters today owe their existence to the Amchitka diaspora of the 1960s and early 1970s. Thus as threatening as atomic tests were to North Pacific wildlife and ecology, *Enhydra lutris* may be less well off had nuclear damage at the island not occurred. Cold War science and policy energized marine mammal conservation in crucial and long-lasting ways.

While fleeing Amchitka benefited species recovery, the island itself and the sea otters that died there also contributed to scientific breakthroughs. Much of wildlife biologist Karl Kenyon's work on the sea otter for the U.S. Fish and Wildlife Service took place at Amchitka in the early 1960s. Kenyon's research became the seminal 1969 study, *The Sea Otter in the Eastern Pacific Ocean*.[18] James Estes arrived at the island in 1970 to survey the effect of Shot Cannikin on the animals for the Atomic Energy Commission. The research that he conducted there and elsewhere in the Aleutians became crucial for understanding the role of otter predation in shaping kelp forests and nearshore ecosystems. Knowledge of keystone species and the trophic cascade was encouraged by circumstances surrounding the bomb.[19]

In an attempt to respond to criticism during the Amchitka debates and build goodwill among Alaska citizens, the Atomic Energy Commission produced a documentary on sea otter translocation in 1969, one of the first films to showcase the animals. The roughly fourteen-minute short, titled *The Warm Coat*, discusses the fur trade, early conservation in the Aleutians, and the growth of the Amchitka population. Along with historical information, producers utilized levity and anthropomorphism to engage their audience. "In the sea, he has grace and style," declares the narrator, and "it's good to relax after a busy day" as imagery of frolicking sea otters is showcased. Additionally, *The Warm Coat* includes an early attempt to personalize the sea otter for a popular audience, as it features the antics and experiences of an ostensibly fictionalized translocation representative dubbed Harvey.[20] (Highlighting the experiences of individual otters for conservation purposes was also practiced by zoo officials, as noted below.) Ultimately, Atomic Energy Commission public relations campaigning may have worked to quell some of the antitesting sentiments of Alaska officials, but it did little to stop the wider controversy surrounding Amchitka.[21] Translocation efforts, as important as they were for marine mammal recovery in the Pacific, never silenced the environmental protests. After the Cannikin blast, further testing plans at Amchitka were canceled.

From "Sea Otter" to "Milk" Island

The geopolitical dynamics of World War II and the postwar period affected sea otters in the western Pacific. The Kuril Islands came under

Soviet control in 1945, and soon thereafter officials were stressing the value of the archipelago's natural resources. Fishing and whaling industries were key to Russian economic integration of the new territories.[22] Yet decades of marine mammal conservation success at the Commander Islands also followed in the wake of the Red Army, with measurable benefits for Kuril otters. Japanese commercial harvests prior to the war may have slowed population recovery, but the USSR's policies provided for notable growth through the rest of the twentieth century. For example, researchers recorded two to three hundred otters at Iturup Island in the 1970s. The number of animals at the island was over a thousand by 1991.[23] Whereas the advance of promyshlenniki through the Kurils in the eighteenth century posed a direct threat to local herds and to Japanese and Ainu economies that relied on them, Soviet sovereignty brought some environmental success to the islands. Fishermen and collective farmers replaced fur traders. The return to Urup Island, once known for its lucrative otter herds, did not result in damaging pelt harvests. Instead, locals in the late twentieth century dubbed it "Milk Island" for its abundant dairy production.[24]

Russians may have seen themselves as protecting Kuril sea otters from capitalist marauders, but other aquatic animals suffered as a result of the takeover. Various whale species were dramatically overhunted by Soviet ships operating from island stations beginning in the late 1940s, first in western Pacific waters and later off North American coasts. As Yulia Ivashchenko and Phil Clapham discuss, "In the space of just 2 years, 1962–63, three new, large whaling factory ships were added to the Soviet North Pacific whaling operation, with the main focus remaining in the eastern North Pacific. As a result of this expansion, catches—many of them illegal—dramatically increased. . . . Catches of sperm whales (the primary target of Soviet whalers in the North Pacific) increased five-fold from 1962 (3,035) to 1966 (15,205). Such intensive whaling continued in the North Pacific until 1969, with up to four Soviet whaling fleets working in the area simultaneously."[25]

The slaughter may be related to killer whale assaults on sea otters later in the century, as discussed below. Whatever its ecological ramifications, the stunning loss of cetacean life also points backward in otter history to where it began. Just as Russian fur hunters moved eastward across the Pacific in the eighteenth century, Soviet whalers pursued their quarry first in its western portions. Sea otters thus share with depleted whale species a geographic kinship that signals the importance of maritime developments emanating from the Pacific West.

Friends and Enemies

As translocations were under way in the eastern Pacific, other factors converged to assist otter conservation in the late twentieth century. American television in the 1970s played an important role in highlighting the species for mass audiences. Naturalist Jacques Cousteau dedicated an episode of his *The Undersea World of Jacques Cousteau* series to the sea otter in 1971. *Mutual of Omaha's Wild Kingdom* followed suit in 1973.[26] These mainstream nature documentaries included little in the way of aesthetic indulgence, although Cousteau could not resist a brief anthropomorphic voice-over: "Otter, with your bristling silver whiskers, you look most wise to me." As the broadcast reported, the gradual expansion of sea otters along California shores was leading to increasing tensions with commercial abalone fishermen by the early 1970s. Sea otter cuteness had a role to play in the debates, as noticed by a central coast columnist in 1970 who quipped that people "actually cooed" when looking at pictures of the animals in a biologist's office and made the following admission: "One of the characters in the story is one of the most appealing and formerly one of the most abused creatures of the sea[, which] makes it such an emotional issue that it is hard to get down to the facts and view the situation dispassionately."[27]

Indeed, enthusiastic captivation with *Enhydra lutris* was more pronounced in California than it was in Alaska due in part to the relative difficulty the species had rebounding in the Golden State, its rarity helping to elicit emotional responses and inspire activism. The nongovernmental organization Friends of the Sea Otter was also crucial in cultivating sentimental attachment by engaging in a vigorous local advocacy campaign. The group was founded in 1968 in the Monterey

Bay area as clashes between abalone fishermen and environmentalists were beginning to intensify. Margaret Owings, a Big Sur resident and nature lover, drafted a column for the *Monterey Peninsula Herald* in March of that year in which she criticized the arguments being made by fishing interests. According to Owings, sea otters were not vicious destroyers of abalone, as they had been portrayed. As she contended, the animals were still very much threatened by poachers, by a potential return of the fur industry, and by oil and other pollutants. Published under the headline "Do Sea Otters Have Any Friends?," the column inspired other Monterey area residents to join with Owings and found the first organization dedicated to protecting the marine mammals.[28]

Friends of the Sea Otter and their supporters relied to a great deal on the animal's aesthetic qualities to promote conservation while boosting local tourism and their organization's profile. A 1972 newspaper article dubbed sea otters "Teddy Bears of the Ocean" and highlighted a trip organized by Owings that brought 375 people together to view Monterey Bay wildlife at close range: "Four boats made two trips each from Fishermen's Wharf and each spent about an hour among the otters while cameras captured their antics and visitors from as far away as New York and Florida 'oh'd' and 'ah'd' in admiration."[29] A *Sports Illustrated* report in 1976 on the controversies with shellfishermen took note of Friends of the Sea Otter's "feisty" newsletter the *Otter Raft*, which, as the columnist noted, included testimonials from concerned correspondents "along with a lot of cute otter illustrations, ads for otter books, posters and movies, [and] fund raising appeals."[30] Overall, public attitudes toward otters evolved during the early to mid-1970s just as political currents carried conflicts between California fishermen and nature defenders in new federal directions.[31]

In conjunction with local, state, and national environmental advocacy, sea otter legal protection advanced in substantial ways in the last quarter of the twentieth century. Like the 1911 statute that it helped to replace, the North Pacific Fur Seal Act of 1966 included bans on taking northern fur seals and otters but applied only to international waters. Passage of the Marine Mammal Protection Act in 1972 and the Endangered

Species Act in 1973 represented "watershed US Congressional actions with unequivocal benefits to sea otter conservation," according to VanBlaricom.[32] The Marine Mammal Protection Act prohibits hunting, killing, or harassing marine mammals within U.S. waters and defines species that fall below an "optimum sustainable population" to be "depleted" and worthy of additional protections. The Endangered Species Act allows animals to be listed as "endangered" or "threatened" even if their current population size is not diminished.[33] By 1977 the California sea otter was listed as threatened under the Endangered Species Act due to its sluggish growth and a perceived threat from offshore oil development. A controversial provision of the Marine Mammal Protection Act allows for subsistence use of marine mammals by Alaskan Natives for creating and selling clothing and handicrafts produced from skins. A Fish and Wildlife Service regulation implemented in 1974 requires that articles be "significantly altered from their natural form" in order to be sold to a non-Native, a vague and confusing statute that has led to confiscations and legal battles since.[34]

The effect of the Native Alaskan industry on otter populations has been a concern for conservationists in recent years. According to Brenda E. Ballachey and James L. Bodkin, continuation of the modern-day harvest, which is concentrated in central and southeastern Alaskan communities, will likely not produce "large-scale consequences for sea otter conservation . . . [but] could either serve to moderate rates of population growth and expansion or, at localized scales, lead to reduced abundance and distribution of sea otters."[35] Public wariness regarding the hunting provision of the Marine Mammal Protection Act has been exemplified in recent press accounts of hunter and artist Peter Williams. Part Yup'ik Eskimo, Williams legally kills sea otters in the waters around Sitka and sews their furs into items such as hats, earrings, and pillows, which he sells online and exhibits at fashion shows. According to reports, he views his small one-man business as an opportunity to overcome poverty and substance abuse, which disproportionately plague Alaska Natives. (Williams spent much of his youth in therapy struggling with depression following the death of his father and four brothers from alcohol-related incidents.) Despite often vocal criticism, Williams draws both economic and spiritual benefits from sustainably hunting otters and sees many of the federal regulations he faces as a new kind of colonialism. The title of one article on his work reflects the popular tension surrounding what he and other hunters do: "Why Would Anyone Want to Shoot a Sea Otter?"[36]

Canadian fisheries regulations were extended to protect the sea otter beginning in 1970, and within the decade the species was listed as "rare" under British Columbia wildlife law. As it returned to unoccupied habitat on the western coast of Vancouver Island it expanded rapidly, more so than was the case along the central mainland coast of western Canada.[37] Yet translocations were done without consulting First Nations peoples. Thus the ecological changes introduced by foraging sea otters have forced many communities to reevaluate fishery preferences. Kelp-associated fish are increasingly available in British Columbia today, but valuable shellfish are less abundant. Some biologists speculate that illegal killing of otters and entanglement in nets are factors responsible for lagging recovery on Vancouver Island.[38] Thus in both California and British Columbia (as well as in southeastern Alaska), sea otters competed with human groups for marine resources as they recolonized lost territory—much as they had for thousands of years in the Pacific prior to the widespread depletions of the fur trade.[39] In different localities one result has been a perception of them as vermin that requires elimination as opposed to unrestrained range expansion.

Aquaria, Oil, and Orcas

While the first efforts in the 1950s to translocate Amchitka sea otters to former coastal habitats in the Pacific failed, U.S. wildlife managers were catching up with Russian scientists in keeping the animals in captivity. The first representatives held for public display at a North American facility were sent from Amchitka to the Woodland Park Zoo in Seattle in 1954. The otters were soon transported to the National Zoological Park in Washington DC, where all died within ten days. As the biologists involved noted in 1955, "Had we known about the success of the Russian work in the [Commander Islands], we might have benefited from their experience in keeping confined otters in good health, particularly as regards the amount of food needed daily."[40]

Yet other otters were successfully kept at the Woodland Park facility from the mid-1950s to the early 1960s. Chief among these was Susie, who was captured at Amchitka by Kenyon and flown to Seattle in October 1955. A report on her death six years later exaggerated her legacy as "the only Sea Otter kept successfully in captivity," yet it made note of Susie's playful nature and highlighted her exhibitory value: "Susie was not only a superb entertainer, but also an excellent subject for students of animal behavior. She was a most interesting and energetic animal, and seemed to be constantly in motion, either swimming, grooming, or feeding. She retained in captivity the Sea Otters' intriguing ability to use a stone as a tool upon which to break the hard shells of clams, etc. . . . The usual method of feeding was always a source of great amazement to our visitors."[41] Building on these early achievements in Washington State, zoos and aquaria in the United States played a crucial role in advancing sea otter research and in generating public support for conservation by the end of the twentieth century. Foremost in these endeavors was the Monterey Bay Aquarium, founded in 1984. The facility developed from associations between the family of David Packard (of Hewlett-Packard fame), local university professors, and Stanford University's Hopkins Marine Station, the last of which sold the deteriorating Hovden Cannery building in Monterey to a nonprofit organization established by the Packards in 1978. The opening of the aquarium six years later was a large celebration heralded by the declaration "The Fish Are Back!," a reference to the city's former sardine industry.[42] As they had with other exhibits, planners and staff met a variety of challenges in attempting to display sea otters for public viewing. The first representatives scratched the acrylic panels of the aquarium tank with shellfish shards, and toys had to be introduced to prevent boredom. The tank was renovated in 1993 with new algae-covered rockwork in an effort to create a more natural enclosure.[43]

Sea otters became star attractions for the internationally recognized Monterey Bay Aquarium, "[the] closest thing to a 'celebrity' species,"

FIG. 13. A sea otter at play at the Monterey Bay Aquarium in 2006. Rocky Yeh, Wikimedia Commons.

according to Connie Chiang.[44] Visitors today are greeted with all manner of stuffed toys, books, and children's material in the facility's gift shops. Outside along Monterey's Cannery Row seaside district, stores sell a myriad of shirts and memorabilia depicting the animals to tourists, evidence of their status as cute and furry emblems of the central coast. Fostering the merchandising of the sea otter allows the aquarium's upper-middle-class patrons to express green values through consumer culture. Yet the degree to which exhibit spectacle and popular consumption produce a beneficial perception of sea otters instead of blurring and trivializing their environmental challenges has been questioned. The many contributions of zoos and aquaria to marine mammal science and conservation largely offset such cultural dilemmas. Nevertheless, officials have recognized some of the inherent problems with exhibiting otters. An article from a 1987 issue of the Monterey Bay Aquarium's newsletter discussing their otter rehabilitation program begins with the question, "Are sea otters too cute and lovable for their own good?," and it warns readers not to try and take home abandoned pups found on area beaches: "People who enjoy the antics of the otters in the aquarium may have the idea that otters in the wild would make ideal pets."[45]

It was also during the 1980s that sea otters grew in popularity in Japan—politically cut off from the animals north of Hokkaido—as a result of aquarium exhibits. In the first half of the decade, the Toba Aquarium obtained four otters, and attendance figures reportedly jumped by 100 percent, spurred on by the birth of a baby otter there in 1984. Accordingly, over a dozen more facilities acquired sea otters by 1987, a growth trend that continued well into the 1990s. By the end of the century over one hundred individuals were being exhibited in Japanese tanks.[46] Despite the popularity of the species, recent years have witnessed a decline in the number of captive sea otters in Japan. Reports cite the difficulty of obtaining new animals due to Japanese restrictions and the poor performance of aquarium breeding programs (due at least in part to the aging of the captive population). As of 2016 only fourteen sea otters remain on display in the country.[47]

With the boom in popular recognition at the end of the twentieth century came a tragic moment, which in turn led to an increase in the number of otters being displayed in aquarium tanks. In 1989 as many as three thousand sea otters died when the *Exxon Valdez* struck a reef in Prince William Sound in Alaska, spilling hundreds of thousands of barrels of crude oil. The otter population of the sound is believed to have fully recovered today, yet the effects of the disaster lingered for decades, as an estimated nine hundred animals died from "chronic exposure or long-term effects of acute exposure," according to scientists.[48] Wildlife death caused by the ship fueled national outrage, especially in regard to *Enhydra*. The reputation that the sea otter had gained by the 1980s allowed it to be pictured in media coverage as the iconic victim of the environmental tragedy. As a columnist for the *Chicago Tribune* noted, "If one image stands out in the aftermath of America's worst oil spill, it is of these once-healthy animals rubbing their eyes and grooming their fur in a futile attempt to rid their coats of slimy crude oil."[49] Although thousands more seabirds died from exposure to the *Exxon Valdez*'s cargo, the plight of sea otters and subsequent efforts to rescue them received the most attention. B. T. Batten notes, "Small, furry, childlike, and vulnerable, sea otters became compelling victims with whom everyone could identify, and thereby made the perfect universal symbol for the injured party. Nearly a year after the spill, a national magazine summarized this sentiment when it referred to a Federal indictment as 'the case of *Otter et al. v. Exxon.*'"[50] Thus sea otter aesthetics, despite problematic concerns, worked to focus attention on the crisis in Alaska while building public support for the species.

Thousands of individuals may have perished following the poisoning of Prince William Sound, but a much deadlier enemy lurked beneath Alaskan waters. Beginning around the time of the *Exxon Valdez* event and continuing into the early twenty-first century, sea otter numbers declined dramatically in the Aleutian Islands and the southern Alaska Peninsula. Approximately eighty thousand animals throughout the region were lost, a death rate rivaling—perhaps even surpassing—the intensity of the Russian hunt of the last half of the eighteenth century. By 2005 sea otters in southwestern Alaska were listed as threatened under the Endangered Species Act.[51] The likely culprits of the massacre were killer whales (*Orcinus orca*), and although some scientists question the circumstantial nature of the evidence against the whales, it makes a compelling case. For example, while there were few sightings of orca attacks on sea otters prior to the 1990s, the number of such reports increased beginning at that time. Additionally, beached carcasses were not being discovered in the affected areas at either a normal rate or one that would correspond with increased death from disease or starvation, suggesting that something was eating the otters. Sea otter behavior in the Aleutians also reportedly changed, as the animals were spending more time closer to shorelines and concentrating in shallow, protected locales.⁵² Just as the maritime fur trade may have influenced the hauling-out habits of the animals, their instincts to flee orcas and modify their use of nearshore spaces reveal something of the agency of *Enhydra lutris* in the face of dangers.

Determining why some killer whales began preying upon sea otters has probably been more controversial than identifying them as a cause of the Alaska decline. Estes and others have argued for a "megafaunal collapse hypothesis," linking the peak of industrial whaling in the North Pacific after World War II with subsequent declines in sea lion and sea otter populations. As whales became less readily available in the region, so goes the argument, orcas turned to other marine mammals for food during the last few decades of the century. This has been subject to much debate among whale and pinniped researchers and ecologists over the last decade and a half. In a memoir published in 2016, Estes summarized the controversies and pointed to economic and political factors as to why megafaunal collapse remains such a contentious issue. "The . . . hypothesis would absolve the ground fishery of ecological wrongdoing or even lead to the management of killer whales," Estes writes. "And some of it may have been for worry by the established research community over what it would mean for their future funding."⁵³

The Trouble with Cuteness

Cuteness is a less intense contemporary concern among marine mammal specialists but one no less worthy of examining. General audience books about sea otters have been some of the more influential cultural artifacts associated with the animals in recent decades. Often large and featuring stunning color photographs of otters in the wild, these volumes have helped to promote the species and shape its reception in crucial ways. There are two notable popular natural histories, one written by California journalist Roy Nickerson and one by VanBlaricom. Nickerson's *Sea Otters: A Natural History and Guide* (with photographs by Richard Bucich) is filled mostly with close-up shots of sea otter faces and captions that periodically evoke the expressiveness of the individuals in the photos: "This resting sea otter does not look happy at being disturbed"; "I can still touch my toes."⁵⁴ VanBlaricom's *Sea Otters* includes a wider variety of images and a more clinical approach to captioning, yet he cannot resist casting his subject in an anthropomorphic light:

> One day I stood with a colleague on the public pier at the town of Pismo Beach, watching sea otters feed on Pismo clams. . . . We saw a young male otter with one clam securely tucked into an axilla [armpit]. He dove repeatedly, but could not find a second clam. Periodically he would stop to rest, bring the clam to his mouth, bite at the unforgiving shell and, in frustration, even pound the clam against his chest. Again and again he would dive, and again and again he would resolutely probe the equally stubborn clam in hopes of breaking through. Finally, just as the sun dropped into the western sea, the otter gave in. Resting quietly on the surface with his gaze averted, he released his grip and allowed the clam to slide from his chest. In the growing darkness he moved off to the north after an embarrassed glance in our direction.⁵⁵

Numerous children's books have been devoted to sea otters. Nonfiction titles regularly employ the "teddy bear" analogy in their text. Equating the marine mammals with stuffed toys complements many charming photographs and affirms the cuddly identity of otters in the minds of young readers: "Certain animals inspire our curiosity and love. Sea otters are like that. Perhaps it's their teddy-bear faces."⁵⁶ "To many people, sea otters look a little like floating teddy bears. But they are much bigger—sea otters can grow to be about 100 pounds."⁵⁷ "[Chapter 1:] Teddy Bear of the Sea."⁵⁸

One problem with such expressions is that the sea otter and its charismatic profile can at times be regarded as a cultural punchline, a minimizing of the animal that has both social and political consequences. In 2013 Alaska state senator Bert Stedman proposed a $100 bounty on otters to incentivize Native Alaskan hunters to help alleviate pressures on southeastern Alaska fishermen. As a report on the proposal notes: "'They're cute and cuddly guys,' says Stedman. 'But their impact, when you just let them overpopulate, on the human side of the equation, is substantial.'"[59] Elsewhere Stedman was quoted as saying, "'This is serious. I try to put humor in it, because we all recognize they are cute and cuddly animals in the water. It sounds Draconian at first but when you take a look at the impact of coastal Alaska, it's a whole different outlook.'"[60] "They're cute . . . but" suggests that the aesthetic appeal of sea otters is what is most important about them, a conception that complicates efforts to advocate for the animals. Aquatic teddy bears prove resistant to the sober attention that the species needs.

The point here is not that the benefits of sea otter cuteness are outweighed by this cultural conundrum. In a 2011 report researchers Sadie S. Stevens, John F. Organ, and Thomas L. Serfass discussed the usefulness of sea otters as a "flagship species." According to the authors, "Flagship species are defined as 'popular, charismatic species that serve as symbols and rallying points to stimulate conservation awareness and action.'"[61] They note the value of otter tourism in California and point to the viral success of *Otters Holding Hands.* No doubt a flagship candidacy for *Enhydra lutris* is secure, and its popularity has done much to raise awareness regarding its environmental needs in recent years. Still, future research should continue to critically assess both the productive and counterproductive aspects of sea otter cuteness and charisma. Children's books, Internet memes, and contemporary films like *Finding Dory* all provide rich ground for analysis.

One way that biologists and others have attempted to mitigate the excesses of sea otter cuteness is through emphasizing the "wild" characteristics of the species. Interviewed by the *San Francisco Chronicle* in 1994, Marianne Riedman, the Monterey Bay Aquarium's chief sea otter researcher at the time, contended, "No doubt they're cute and fun to watch, but remember, they are, after all, wild carnivorous animals."[62] Among less appealing facts about otters, Riedman noted that the creatures often steal food from each other and that females are commonly wounded during mating, sometimes with lethal injuries. The animals "are immensely powerful—so strong, Riedman says, that it would take three strong men to hold one down."[63] While photos of females with mating wounds on their noses are displayed in books and nature documentaries, the image of sea otters as aggressive and even sinister has found a degree of ironic "clickbait" resonance in recent Internet media. Author Brian Switek reported on research that analyzed sea otters in Monterey Bay attempting to mate with harbor seal pups even after the violent encounters led to the deaths of the young seals. "At least two of the sea otters had been previously held at the Monterey Bay Aquarium as part of their rehabilitation program," writes Switek, explaining that "the trouble they experienced early in their lives might have made them more likely assailants."[64]

Such information becoming more widely known beyond marine mammal science might work to complicate the animals in the public imagination and promote more nuanced understandings of them. Larry Pynn of the *Vancouver Sun* detailed the "dark side" of sea otter behavior by recounting the story of Whiskers, a Vancouver Island sea otter who repeatedly attempted to entice dogs into the surf until one day he was found trying to mate with a dead canine in the water. Pynn writes, "Cute and cuddly—not rapist and murderer—are mentioned in the typical bio on sea otters."[65] Mirroring Riedman's corrective, Vancouver Aquarium marine mammal curator Brian Sheehan is quoted in the 2014 story as saying, "That's one thing we have with our visitors: 'Oh, they look adorable like stuffed animals,' . . . [b]ut they definitely have the potential for being strong and aggressive." Nevertheless, as poignant as Switek's commentary may be that "the dark side of superficially cute animals is a part of their nature that reminds us that the wild does not exist for our entertainment and whimsy," how this less sentimental appreciation for the species can gain traction in a world of fuzzy toys, tourist sweaters, and otter memes is difficult to envision. Even the photos accompanying Pynn's article that capture the animals engaged in instances of odd sexual behavior do not appear overtly sinister—in a couple of the images, a sea otter seems to be playing with, not attempting to molest, a cormorant seabird.

Otters Holding Hands remains the most widely viewed sea otter video on YouTube. As of 2017 Holmes's clip has been viewed over twenty-one million times. She is currently a business professor at a Canadian university and is often contacted by production companies in Japan capitalizing on nationally popular Internet clip shows and Japanese cuteness traditions (*kawaii*).[66] According to Holmes, she donates money for licensing *Otters Holding Hands* to the Vancouver Aquarium, and while she is sympathetic with those who question the commercialization of nature, she remains proud of her creation and her connection to Nyac's and Milo's lives. "The video has become a proxy for the animals," Holmes contends, and it has helped to promote conservation and build deeper human-animal relationships. "Some individual otters have become 'spokesotters,' and they are doing this behavior that humans can relate to, this thing that looks like affection and might even be affection in some way."[67]

Nyac and Milo have since died at the Vancouver Aquarium, but poet Dina Del Bucchia produced eulogies for them that provide an alternative way to remember the two famous sea otters. She expresses the inherent tension of cuteness and prods us to question our Internet browsing histories. For Nyac, Del Bucchia wrote:

> From slicked black, snout not even visible,
> to picture-perfect, made for advertisements, plush toys, mugs.
> Fur-print tote bags instead of torn from your flesh.
> You had the right story, a TV movie starring
> Jennifer Love Hewitt, that you overcame with
> take-a-look-at-me-now appeal. You were a girl fished
> from a well, a kidnapping survivor, a wartorn orphan,
> a slim pup reborn in oil.[68]

For Milo:

> Like an ailing politician, we weep
> for Milo, hold vigils
> on the Internet. I advise
> we make small shrines in our homes:
> tasteful glitter pens, foam core,
> fake candles, ceramic replicas
> looted from the gift shop.[69]

Conclusion

Sea otters have returned to Glacier Bay in southeastern Alaska after an absence that can only partly be blamed on humans. In fact, the bay did not exist prior to the eighteenth century. Glacial ice advanced up to 1750 and covered the surrounding area to the coast. Climate change then set in, causing the ice to recede just as the North Pacific fur rush began. In the late nineteenth century, the naturalist John Muir visited the newly carved bay and recorded the glacial retreat, as did the Harriman expedition, which also noted the scarcity of Alaskan otters. President Calvin Coolidge proclaimed Glacier Bay a national monument in 1925. Later expansions led to the Glacier Bay National Park and Preserve in 1980. By that same decade, translocation efforts eventually succeeded as sea otters began recolonizing the area. Fueled by the bay's natural resources, historic growth rates occurred at the turn of the millennium. As conservation biologists Perry Williams and Melvin Hooten summarize, "Today, they are one of the most abundant marine mammals in Glacier Bay. Recent observations have documented large groups of more than 500 sea otters in some parts of lower Glacier Bay, suggesting that prey resources are abundant."[1]

Animals colonizing a geologically recent Pacific formation is a unique event in many ways, but like the other actors in this book, the sea otters of Glacier Bay symbolize the dynamic and often dramatic forces that have influenced the lives of the creatures in recent centuries. They are social beings linked across time to kelps, urchins, sea cows, and beavers. Their skins have kept human bodies and their own warm and comfortable. Global trade networks and commercial enterprises nearly destroyed them. Science and diplomacy helped to restore them. Sea otters are more than cute icons. They have history.

Knowing more about that history will provide greater knowledge of both sea otters and their Pacific World. It will further illuminate how empires contended for territory and connected the ocean, how the fur trade changed the Pacific's places, people, animals, and plants. Telling new sea otter stories will illustrate ecological change across time and space, how otters are intimately linked to other living things in and near their marine homes. Lastly, sea otter history will hopefully continue to improve our relations with *Enhydra lutris*, to share in and appreciate the species' complex past and, as in an aquarium visit, connect with it in the present. There are historical gaps that need to be filled. This book has hopefully inspired some of the work that remains.

APPENDIX

List of Vessels Engaged in the California Sea Otter Trade, 1786–1847

The historian Adele Ogden included a list of California sea otter trade vessels in her 1941 book *The California Sea Otter Trade, 1784–1848*. Ogden's knowledge of maritime history expanded over the course of her career, culminating in a 1979 manuscript titled "Trading Vessels on the California Coast, 1786–1848," which she donated to the Bancroft Library.[1] The data presented here are a synthesis of both sources and a revised list of vessels engaged in the sea otter trade. As Ogden understood, maritime records provide an incomplete picture of the California trade, yet a full presentation of the information she compiled can benefit future researchers. An earlier attempt by historian Glenn Farris to reproduce her 1979 manuscript included additional data that she did not and was only a partial list of vessels and pelt exports.[2] This appendix is an attempt at a complete compilation of Ogden's two sources and is intended as a commemoration of her groundbreaking research in California history.[3]

The format for listing vessels here generally follows Ogden's original 1941 list but only includes information relating to sea otter skins shipped out of California, the years a particular ship arrived at and departed from the coast, and the ship's national origin. Data relating to ship personnel or voyage records beyond California are omitted. Supplemental descriptions of sea otter skins shipments that Ogden provided in either 1941 or 1979 are included. Parenthetical notations for a lack of information on otter cargoes are not from either of her lists. A graph for the data illustrating the volume of sea otter pelts taken from California by decade is also provided. It is modeled after a graph made from Ogden's 1941 data that was published by marine mammal researchers in 2011.[4] As comparison of the two graphs suggests, Ogden located additional vessels in later years of her research, particularly for the years after 1820. She also modified cargo data for a number of ships in her 1979 manuscript.

Sea Otter Skins Taken Each Year and Annual Totals

Year departed	Skins (taking average value if a range is given)	Year departed	Total skins
1786	1,060	1786	1,060
1787	1,750	1787	1,750
1788	116	1788	116
1789	234	1789	234
1790	656	1790	656
1803	491	1803	2,091
1803	1,600	1804	1,800
1804	1,800	1805	292
1805	292	1806	2,444
1806	17	1807	9,784
1806	2,427	1808	0
1807	2,848	1809	6,167
1807	273	1810	276
1807	4,819	1811	9,356
1807	1,231	1812	5,250
1807	613	1813	1,603
1809	2,350	1814	392
1809	2,117	1815	943
1809	1,700	1816	0
1810	96	1817	0
1810	20	1818	72
1810	160	1822	206
1811	1,190	1823	56
1811	2,976	1824	142
1811	3,952	1825	375
1811	1,238	1826	606

Year	Value	Year	Value
1812	1,442	1827	0
1812	1,792	1828	190
1812	1,516	1829	109
1812	500	1830	0
1813	1,603	1831	1,182
1814	392	1832	682
1815	8	1833	740
1816	955	1834	291
1818	72	1835	424
1822	150	1836	68
1823	56	1837	0
1823	46	1838	0
1823	10	1839	5
1824	303	1840	190
1824	14	1841	0
1824	110	1842	58
1824	18	1843	44
1825	375	1844	21
1826	468	1845	89
1827	138	1846	158
1828	40	1847	47
1828	150		
1829	63		
1829	40		
1829	6		
1831	4		
1831	300		
1831	478		
1832	160		
1832	100		
1832	316		
1832	106		
1833	96		

Total: 49,969

1833	350
1833	170
1833	10
1833	114
1834	188
1834	90
1834	8
1834	5
1835	18
1835	6
1835	400
1836	68
1839	5
1840	16
1840	3
1840	155
1842	47
1842	11
1843	44
1844	80
1844	21
1845	39
1845	50
1846	55
1846	103
1847	2
1847	5
1847	40

Source: Adele Ogden, *The California Sea Otter Trade, 1784–1848* (Berkeley: University of California Press, 1941); and Ogden, "Trading Vessels on the California Coast, 1786–1848," Bancroft Library, University of California, Berkeley.

Catch by Period

Year range	Total skins	% of total skins
1786–1800	3,816	7.64
1801–9	22,578	45.18
1810–19	17,892	35.81
1820–29	1,684	3.37
1830–39	3,392	6.79
1840–47	607	1.21
Total	**49,969**	

Source: Adele Ogden, *The California Sea Otter Trade, 1784–1848* (Berkeley: University of California Press, 1941); and Ogden, "Trading Vessels on the California Coast, 1786–1848," Bancroft Library, University of California, Berkeley.

Ships and Catches in the California Sea Otter Trade

Year	Ship	Flag	Skins	Specifics
1786	*Princesa*	Spain	1,060	Sea otter skins
1787	*Favorita*	Spain	1,750	Sea otter skins (joint cargo of Favorita and San Carlos)
1787	*San Carlos*	Spain		Sea otter skins (joint cargo of Favorita and San Carlos)
1788	*Aranzazu*	Spain	116	Sea otter skins (at San Francisco)
1789	*Aranzazu*	Spain	234	Sea otter skins (at Santa Barbara)
1790	*San Carlos*	Spain	656	Sea otter skins
1793	*Butterworth*	England		(No data on otter skins)
1793	*Jackal*	England		(No data on otter skins)
1793	*Prince Lee Boo*	England		(No data on otter skins)
1794	*Butterworth*	England		(No data on otter skins)
1794	*Jenny*	England		(No data on otter skins)

Year	Ship	Flag	Skins	Specifics
1795	*Phoenix*	England		(No data on otter skins)
1796	*Otter*	United States		(No data on otter skins)
1798	*Garland*	United States		(No data on otter skins)
1799	*Eliza*	United States		Sea otter skins (inward cargo)
1800	*Betsy*	United States		(No data on otter skins)
1801	*Enterprise*	United States		(No data on otter skins)
1803	*Alexander*	United States	491	Sea otter skins (onboard at San Diego, confiscated)
1803	*Alexander* (second trip)	United States		(No data on otter skins)
1803	*Hazard*	United States		(No data on otter skins)
1803	*Leila Byrd*	United States	1,600	Sea otter skins (Purchased at San Blas, January 1803)
1803–4	*O'Cain*	United States	1,800	Sea otter skins (700 sold to Californians)
1804	*Hazard*	United States		(No data on otter skins)
1804	*Hazard* (second trip)	United States		(No data on otter skins)
1804	*Leila Byrd*	United States		Furs
1804	*Princesa*	Spain		Seven bundles of sea otter skins
1805	*Activo*	Spain	292	Sea otter skins
1805	*Leila Byrd*	United States		Sea otter skins
1805	*Princesa*	Spain		Three bundles of sea otter skins
1806	*Eclipse*	United States		(No data on otter skins)
1806	*O'Cain*	United States	17	Sea otter skins (valued at $60,000)
1806	*Peacock*	United States		(No data on otter skins)
1806	*Tamana*	United States	2,427	Sea otter skins
1806–7	*Maryland*	United States		(No data on otter skins)

Year	Ship	Flag	Skins	Specifics
1806–7	*Mercury*	United States	2,848	Sea otter skins (1,772 prime, 1,076 small)
1807	*Activo*	Spain	273	Sea otter skins (joint cargo of *Activo* and *Princesa*)
1807	*Derby*	United States		(No data on otter skins)
1807	*O'Cain*	United States	4,819	Sea otter skins (3,006 prime, 1,264 yearlings, 549 cubs)
1807	*Peacock*	United States	1,231	Sea otter skins (753 prime, 228 yearlings, 250 cubs)
1807	*Princesa*	Spain		Sea otter skins (joint cargo with *Activo* and *Princesa*)
1807	*Tamana*	United States	613	Sea otter skins
1808–9	*Kodiak*	Russia	2,350	Sea otter skins (1,453 grown, 406 yearlings, 491 pups)
1808–9	*Mercury*	United States	2,117	Sea otter skins (1,688 grown, 256 yearlings, 136 cubs, 37 others)
1809	*Dromo*	United States	1,700	Sea otter skins
1809	*Isabella*	United States		(No data on otter skins)
1809–10	*Mercury*	United States	96	Sea otter skins (at Santa Barbara)
1810	*Albatross*	United States		(No data on otter skins)
1810	*Mercury*	United States	20	Sea otter skins (at Santa Barbara)
1810	*Princesa*	Spain	160	Sea otter skins
1810–11	*Albatross*	United States	1,190	Sea otter skins (778 grown, 140 yearlings, 202 pups, 70 others)
1810–11	*Isabella*	United States	2,976	Sea otter skins (1,978 grown, 432 yearlings, 566 pups)

Year	Ship	Flag	Skins	Specifics
1810–11	O'Cain	United States	3,952	Sea otter skins
1811	Albatross	United States		(No data on otter skins)
1811	Chirikov	Russia	1,238	Sea otter skins (1,160 grown, 78 yearlings)
1811–12	Chirikov	Russia		(No data on otter skins)
1812	Albatross	United States		(No data on otter skins)
1812	Amethyst	United States	1,442	Sea otter skins (1,310 grown, 98 yearlings, 34 pups)
1812	Charon	United States	1,792	Sea otter skins (1,596 grown, 136 yearlings, 60 pups)
1812	Katherine	United States	1,516	Sea otter skins (1,252 grown, 186 yearlings, 78 pups)
1812	Mercury	United States	500	Sea otter skins
1813	Albatross	United States		(No data on otter skins)
1813	Mercury	United States	1,603	Sea otter skins (947 Sea otter tails, at time of seizure)
1813	O'Cain	United States		(No data on otter skins)
1814	Albatross	United States		(No data on otter skins)
1814	Charon	United States		(No data on otter skins)
1814	Forester	United States		(No data on otter skins)
1814	Ilmen	Russia	392	Sea otter skins (322 grown, 50 yearlings, 20 pups)
1814	Isabella	United States		(No data on otter skins)
1814	O'Cain	United States		(No data on otter skins)
1814	Pedler	United States		(No data on otter skins)
1814–15	Forester	United States		(No data on otter skins)
1815	Forester	United States		(No data on otter skins)
1815	Chirikov	Russia	8	Sea otter skins

Year	Ship	Flag	Skins	Specifics
1815–16	*Ilmen*	Russia	955	Sea otter skins (obtained around Santa Barbara Channel Islands)
1816	*Albatross*	United States		(No data on otter skins)
1816	*Atala*	United States		(No data on otter skins)
1816	*Lydia*	United States		(No data on otter skins)
1816	*O'Cain*	United States		(No data on otter skins)
1816	*Sultan*	United States		(No data on otter skins)
1816–17	*Traveller*	United States		(No data on otter skins)
1817	*Avon*	United States		(No data on otter skins)
1817	*Bordeaux Packet*	United States		(No data on otter skins)
1817	*Bordelais*	France		Sea otter skins
1817	*Bordelais* (second trip)	France		(No data on otter skins)
1817	*Chirikov*	Russia		(No data on otter skins)
1817	*Columbia*	England		(No data on otter skins)
1817	*Cossack*	United States		(No data on otter skins)
1817	*Kutusov*	Russia		(No data on otter skins)
1818	*Bordelais*	France		(No data on otter skins)
1818	*Clarion*	United States		Sea otter skins
1818	*Eagle*	United States		(No data on otter skins)
1818	*Kutusov*	Russia	72	Sea otter skins (at Santa Cruz)
1818	*Okhotsk*	Russia		(No data on otter skins)
1818	*Ship*	United States		(No data on otter skins)
1820	*Ilmen*	Russia		(No data on otter skins)
1820	*San Francisco de Paula*	Spain		(No data on otter skins)
1821	*Eagle*	United States		Sea otter skins
1822	*Eagle*	United States		(No data on otter skins)
1822	*Owhyhee*	United States	150	Sea otter skins

Year	Ship	Flag	Skins	Specifics
1822–23	*Volga*	Russia	56	Sea otter skins (40 grown, 16 yearling)
1822–23	*Sachem*	United States		Sea otter skins
1823	*Ann*	United States		(No data on otter skins)
1823	*Buldakov*	Russia	46	Sea otter skins (44 grown, 2 yearlings)
1823	*John Begg*	England		Sea otter skins
1823	*Mentor*	United States		Sea otter skins
1823	*Rover*	United States	10	Sea otter skins
1823–24	*Rover*	United States	303	Sea otter skins (300 sea otter tails)
1824	*Becket*	Hawaii		(No data on otter skins)
1824	*Mentor*	United States	14	Sea otter skins
1824	*Owhyhee*	United States	110	Sea otter skins
1824	*Owhyhee* (second trip)	United States		(No data on otter skins)
1824	*Sultan*	United States		(No data on otter skins)
1824	*Washington*	United States	18	Sea otter skins
1824–25	*Rover*	Mexico	444	Sea otter skins (375 sea otter tails, 69 sea otter skins and pieces)
1825	*Nile*	United States		(No data on otter skins)
1825	*Tamaahmaah*	United States		(No data on otter skins)
1825	*Washington*	United States		(No data on otter skins)
1825–26	*Baikal*	Russia	468	Sea otter skins
1826	*Convoy*	United States		(No data on otter skins)
1826	*Harbinger*	United States		(No data on otter skins)
1826	*Owhyhee*	United States		(No data on otter skins)
1826	*Rover*	United States		(No data on otter skins)
1826	*Washington*	United States		(No data on otter skins)
1826–27	*Waverly*	Hawaii	138	Sea otter skins
1826–27	*Harbinger*	United States		(No data on otter skins)

Year	Ship	Flag	Skins	Specifics
1827	*Kamehameha*	United States		(No data on otter skins)
1827	*Karimoku*	Hawaii		(No data on otter skins)
1827	*Tamaahmaah*	United States		(No data on otter skins)
1827–28	*Harbinger*	United States		(No data on otter skins)
1827–28	*Waverly*	Hawaii		(No data on otter skins)
1828	*Griffon*	United States	40	Sea otter skins
1828	*Héros*	France	150	Sea otter skins
1828	*Karimoku*	Hawaii		(No data on otter skins)
1828–29	*Karimoku* (second trip)	Hawaii		(No data on otter skins)
1828–29	*Baikal*	Russia	63	Sea otter skins
1828–29	*Washington*	United States		(No data on otter skins)
1828–29	*Waverly*	Hawaii		One barrel of sea otter skins
1829	*Dhualle*	England	40	Sea otter skins
1829	*Diana*	United States		(No data on otter skins)
1829	*Volunteer*	United States		(No data on otter skins)
1829	*Washington*	United States	6	Sea otter skins
1829–30	*Brookline*	United States		Sea otter skins
1830	*Volunteer*	United States		(No data on otter skins)
1831	*Convoy*	United States	4	Sea otter skins
1831	*Griffon*	England	300	Sea otter skins
1831	*Louisa*	United States		Sea otter skins (400 sea otter skins, all probably from Sitka)
1831	*William Little*	England	478	Sea otter skins (shipped by Alfred Robinson)
1832	*Crusader*	United States	160	Sea otter skins
1832	*Griffon*	United States	100	Sea otter skins
1832	*Plant*	United States	316	Sea otter skins
1832	*Victoria*	United States	106	Sea otter skins
1832–33	*Crusader*	United States	96	Sea otter skins

Year	Ship	Flag	Skins	Specifics
1833	*Convoy*	United States	300–400	Sea otter skins
1833	*Convoy* (second trip)	United States	170	Sea otter skins
1833	*Harriet Blanchard*	United States	10	Sea otter skins
1833	*Maraquita*	Hawaii	114	Sea otter skins
1833–34	*Loriot*	United States	188	Sea otter skins (165 sea otter tails)
1833–34	*Volunteer*	United States	80–100	Sea otter skins
1833–35	*Lagoda*	United States	18	Sea otter skins
1833–34	*Leonor*	Mexico	8	Sea otter skins
1834	*Bolivar Liberator*	United States		(No data on otter skins)
1834	*Don Quixote*	United States	5	Sea otter skins
1834	*Avon*	United States		Sea otter skins
1834–35	*California*	United States		Sea otter skins
1834	*Convoy*	United States		Sea otter skins and sea otter tails
1835	*Avon*	United States	6	Sea otter skins
1835	*Bolivar Liberator*	United States	400	Sea otter skins
1835–36	*Alert*	United States		Sea otter skins
1835–36	*Diana*	United States		Sea otter skins
1836	*Convoy*	United States		Sea otter skins
1836	*Don Quixote*	United States		Sea otter skins
1836	*Joseph Peabody*	United States		Furs
1836	*Loriot*	United States	68	Sea otter skins (57 grown, 11 pups)
1836–37	*Clementine*	England		Sea otter skins
1837–38	*Lama*	England		Sea otter skins
1837–38	*Rasselas*	United States		Sea otter skins
1838	*Lama*	England		Furs

Year	Ship	Flag	Skins	Specifics
1838	*Lama* (second trip)	England		Sea otter skins
1838–39	*California*	Mexico	5	Sea otter skins
1839–40	*Morse*	United States		Sea otter skins
1839–41	*Monsoon*	United States		Furs
1839	*Thomas Perkins*	United States		Sea otter skins ($9,000 value at Honolulu)
1840	*Alciope*	United States	3	Sea otter skins
1840	*California*	Mexico	16	Sea otter skins (belonging to Henry Delano Fitch)
1840	*Nymph*	Mexico	155	Sea otter skins
1841–42	*Fama*	United States		Sea otter skins
1841–42	*Maryland*	United States	47	Sea otter skins
1841–42	*Don Quixote*	United States		Sea otter skins
1841–44	*Tasso*	United States		Sea otter skins
1842	*Trinidad*	Mexico	11	Sea otter skins
1842–44	*Barnstable*	United States	80	Sea otter skins
1842–44	*California*	United States		(No data on otter skins)
1843–44	*Bolivar Liberator*	United States	21	Sea otter skins
1843–46	*Admittance*	United States	55	Sea otter skins
1843	*Diamond*	England	44	Sea otter skins
1844–45	*Oajaca*	Mexico		(No data on otter skins)
1844–47	*Sterling*	United States	2	Sea otter skins
1844–47	*Vandalia*	United States	5	Sea otter skins (71 sea otter skins carried from San Pedro to Monterey, May 184
1845–46	*California*	United States	103	Sea otter skins
1845	*Don Quixote*	United States	39	Sea otter skins
1845	*Naslednik Aleksandra*	Russia	50	Sea otter skins

Year	Ship	Flag	Skins	Specifics
1845–48	*Tasso*	United States		(No data on otter skins)
1846	*Euphemia*	United States		Sea otter skins
1847	*Loo Choo*	United States	40	Sea otter skins
1847	*Francisca*	United States		Sea otter skis

Source: Adele Ogden, *The California Sea Otter Trade, 1784–1848* (Berkeley: University of California Press, 1941); and Ogden, "Trading Vessels on the California Coast, 1786–1848," Bancroft Library, University of California, Berkeley.

NOTES

INTRODUCTION

1. See Ogden, *California Sea Otter Trade*; Gibson, *Otter Skins*; Scofield, *Hail, Columbia*; Gibson, "Nootka and Nutria"; Malloy, *Boston Men*; Hardee, "Soft Gold."
2. Prominent sources include Igler, "Diseased Goods"; Freeman, *The Pacific*; Matsuda, *Pacific Worlds*; Igler, *The Great Ocean*; Armitage and Bashford, *Pacific Histories*.
3. Igler, "Exploring."
4. For examples, see Zilberstein, "Objects of Distant Exchange"; Farris, "Otter Hunting"; Jones, *Empire of Extinction*.
5. Nance, introduction to *The Historical Animal*, 3.
6. See Jones et al., "Toward a Prehistory."
7. For the earlier version of the chapter, see Ravalli, "Sea Otter Aesthetics."
8. Montanari, "Rare Otter Fossil"; Berta, Sumich, and Kovacs, *Marine Mammals*, 119–21; VanBlaricom and Estes, *The Community Ecology*, 6–7; Boessenecker, "A Middle Pleistocene Sea Otter."
9. Estes, "Natural History," 22. As Estes notes, scientists recognize three subspecies of sea otter based largely on size and morphological differences: *Enhydra lutris lutris* (from the Kuril Islands, Kamchatka, and the Commander Islands), *Enhydra lutris kenyoni* (from Alaska to Washington State), and *Enhydra lutris nereis* (the California otter), from largest to smallest in terms of body size. The divisions correspond roughly with some distinctions that were made regarding pelt quality during the maritime fur trade. Nevertheless, the impact that recognized differences may have had over the larger course of sea otter history appears to have been marginal, but future scholars should challenge this assumption. For the purposes of this book, the sea otter is always referred to generically and without regard for subspecific taxonomy.
10. Jones, *Empire of Extinction*, 9–11, 62–64; Reid, *The Sea Is My County*, 4–7.
11. Estes, "Natural History," 22.
12. Murray, "Veterinary Medicine," 170.
13. Kenyon, *The Sea Otter*, 278–81; Ballachey and Bodkin, "Challenges," 71–76.

14. Bodkin, "Historic and Contemporary Status," 44–45.
15. Larson and Bodkin, "The Conservation of Sea Otters," 6.
16. See Forest, "Searching for Sea Otters."
17. Steller, *De bestiis marinis*.
18. Igler, "The Northeastern Pacific Basin," 588; Igler, "Hardly Pacific." I am adapting Igler's "hardly Pacific" conceptualization for prehistory; he uses it for eighteenth- and nineteenth-century European American maritime activity.
19. Freeman, *The Pacific*, 45–46; Hattori et al., "History and Status"; Yamaura, "The Sea Mammal Hunting Cultures."
20. Erlandson, Moss, and Des Lauriers, "Life on the Edge"; Jones et al., "Toward a Prehistory," 251. For evidence that sea otter hunting increased among the people of Haida Gwaii over the past two thousand years, see Sloan and Dick, *Sea Otters of Haida Gwaii*, 28.
21. Jones, "Running into Whales," 361.
22. Corbett et al., "Aleut Hunters"; Rick et al., "Historical Ecology."
23. Jones, "Running into Whales," 365.
24. Lech, Betts, and Maschner, "An Analysis," 126.
25. Salomon et al., "First Nations Perspectives," 309.
26. McKechme and Wigen, "Toward a Historical Ecology," 155.
27. Moss and Losey, "Native American Use," 185.
28. Salomon et al., "First Nations Perspectives," 308.
29. Salomon et al., "First Nations Perspectives," 308; Fisher, "The Northwest," 122–23; Reid, *The Sea Is My County*, 44–45.
30. Phillipi, *Songs of Gods*, 59.
31. Jones, *Empire of Extinction*, 85.
32. Lech, Betts, and Maschner, "An Analysis," 113. Also see Black, "Animal World."
33. Salomon et al., "First Nations Perspectives," 312–15. For Aleut management of sea otters prior to Russian contact, see Jones, *Empire of Extinction*, 86–89.
34. Sloan and Dick, *Sea Otters of Haida Gwaii*, 20.
35. Koerper, "Two Sea Otter Effigies."

1. RAKKOSHIMA, THE SEA OTTER ISLANDS

1. For a summary of Sato Genrokuro's account of the events at Urup Island, see Walker, *The Conquest of Ainu Lands*, 162–63.
2. Igler, "Diseased Goods," 694–95.
3. Hellyer, "The West."
4. Walker, *The Conquest of Ainu Lands*, 11–12.
5. Pomeranz and Topik, *The World That Trade Created*, 119; Schlesinger, *A World Trimmed with Fur*, 30.

6. Takahashi, "Inter-Asian Competition," 40.
7. Takahashi, "Inter-Asian Competition," 41–43; Hellyer, "The West," 396. For the historic existence of sea otters at Hokkaido, see Hattori et al., "History and Status."
8. Walker, *The Conquest of Ainu Lands*, 20–21; Irish, *Hokkaido*, 23–26.
9. Walker, *The Conquest of Ainu Lands*, 80–81; Tezuka, "Ainu Sea Otter Hunting," 120.
10. Tezuka, "Long Distance Trade Networks," 352.
11. Tezuka, "Long Distance Trade Networks," 355–57.
12. Walker, *The Conquest of Ainu Lands*, 157. For Jesuit missionary reports, see Abé, "The Seventeenth Century Jesuit Missionary Reports."
13. Stephan, *The Kuril Islands*, 7.
14. Walker, *The Conquest of Ainu Lands*, 1–5.
15. Takahashi, "Inter-Asian Competition," 41.
16. See Schlesinger, *A World Trimmed with Fur*, 134, for evidence of an increase in sea otter exports from Japan after 1785.
17. Walker, *The Conquest of Ainu Lands*, 100–109.
18. Barthélemy de Lesseps, *Travels in Kamtschatka*, 208–17. As Katherine Plummer argues in her summary of Japanese sources regarding the incident, the merchants were on their way to deliver rice to the shogun, not to engage in the Kuril trade. See Plummer, *The Shogun's Reluctant Ambassadors*, 43–52.
19. Walker, *The Conquest of Ainu Lands*, 97–98.
20. Walker, *The Conquest of Ainu Lands*, 126.
21. Takahashi, "Inter-Asian Competition," 42.
22. Stephan, *The Kuril Islands*, 66–68.
23. McDougall, *Let the Sea Make a Noise*, 46–54.
24. Stolberg, "Interracial Outposts," 327–28.
25. Stolberg, "Interracial Outposts," 334–36.
26. Gibson, "Sitka–Kyakhta versus Sitka–Canton," 44–45. For other negative assessments of the Russian position in the China trade of the eighteenth century, see Jones, *Empire of Extinction*, 131; Schlesinger, *A World Trimmed with Fur*, 51.
27. Gibson, "Sitka–Kyakhta versus Sitka–Canton," 43. Increasing competition from the Anglo-American sea otter trade at Canton by the end of the eighteenth century, which effectively lowered the prices that Russians could receive at Kiakhta, meant that Chinese merchants returned to their earlier preference for trade in smaller Siberian furs such as squirrel. British captain John Dundas Cochrane, during his journey through the region in the 1820s, noted this preference with curiosity. See Cochrane, *Narrative*, 169.
28. McDougall, *Let the Sea Make a Noise*, 57–58; Jones, *Empire of Extinction*, 28–29.
29. Jones, *Empire of Extinction*, 65–66; Vinkovetsky, *Russian America*, 29–30.

30. When discussing Russian tribute collecting in the northern Kurils in a seminal history of Russo-Japanese relations, George Alexander Lensen mentioned "beaver" as the principal fur sought after by Japanese and Russians alike, yet to the south for a later period he distinguishes sea otters. See Lensen, *The Russian Push*, 32, 65, 69. Merchants from both nations were indeed interested in a variety of furs, but the fact that the northern Kuril Islands are part of the historic range of *Enhydra lutris* supports the likelihood that Lensen erred.
31. Dmytryshyn, Crownhart-Vaughan, and Vaughan, *Russian Penetration*, 45.
32. Krasheninnikov, *The History of Kamtschatka*, 131–32.
33. Walker, *The Conquest of Ainu Lands*, 162; Jones, *Empire of Extinction*, 44–46.
34. Jones, *Empire of Extinction*, 46–49.
35. Steller, *De bestiis marinis*.
36. Black, *Russians in Alaska*, 24. Black notes a mid-1700s prohibition of merchant travel to the Kuril Islands that "affect[ed] the development of the Aleutian trade" (*Russians in Alaska*, 65). Yet as Stephan notes, such directives from the Department of Siberian Affairs (lifted in the 1760s) were not intended for the northern Kurils and were not always heeded by locals (*The Kuril Islands*, 49).
37. Stephan, *The Kuril Islands*, 63n.
38. See Steller, *History of Kamchatka*, 13–17.
39. Stephan, *The Kuril Islands*, 48–50. While Russians did proselytize among and attempt a more humane assimilation of the inhabitants of Shumshu Island (the first south of Kamchatka), the treatment of Native people in the western Pacific at this time differed little from what Aleut Natives experienced during the eighteenth century. Local Japanese were also guilty of exploitation and violent treatment of Ainu.
40. Stephan, *The Kuril Islands*, 61–64; McDougall, *Let the Sea Make a Noise*, 98–102; Wells, *Russian Views*, 3–4.
41. For Shelikhov's earlier attempts at the Kuril trade in partnership with other merchants, see Berkh, *A Chronological History*, 77–78.
42. Stephan, *The Kuril Islands*, 64.
43. Shubin, "Russian Settlements," 429.
44. Stephan, *The Kuril Islands*, 70–71.
45. Walker, *The Conquest of Ainu Lands*, 176.
46. Irish, *Hokkaido*, 66–68.
47. Matthews, *Glorious Misadventures*, 296–99.
48. McDougall, *Let the Sea Make a Noise*, 131–34; Matthews, *Glorious Misadventures*, 299.
49. Stephan, *The Kuril Islands*, 79; Irish, *Hokkaido*, 62–64.
50. Stephan, *The Kuril Islands*, 80–95.

51. Kornev and Korneva, "Population Dynamics," 275; Kornev and Korneva, "Historical Trends," 21.
52. Steller, *De bestiis marinis*. Also see Stephan, *The Kuril Islands*, 98–99.
53. Snow, *In Forbidden Seas*, 291. There was a period of sharp increase of Japanese sea otter exports to China during the 1810s, but it produced on average only about 725 pelts annually and represented at most 15 percent of total Chinese supplies. See Schlesinger, *A World Trimmed with Fur*, 134, 217n26.
54. Tikhmenev, *A History*, 173.
55. Arndt, "Preserving the Future Hunt," 4–6. As evidence of the success of Russian conservation efforts, Tikhmenev notes that a "long closed season in the Urup Island area, where in the early 1840s the sea otters disappeared completely ... remedied the situation and the hunting at this island is now quite good" (*A History*, 357).
56. Kornev and Korneva, "Population Dynamics," 275.

2. PROMYSHLENNIKI AND PADRES

1. Venegas, *A Natural and Civil History*, 222.
2. Torrubia, *The Muscovites*, 44; Saunt, *West of the Revolution*, 52–53.
3. For Spanish uninterest in the fur trade, with the exception of fur and hide trades in New Mexico, see Beals, *Juan Perez*, 27; Mapp, *The Elusive West*, 64–65.
4. Jones, *Empire of Extinction*, 243–53.
5. See Lensink, "The History and Status." For use of Lensink's data in contemporary conservation science, see Bodkin, "Historic and Contemporary Status," 44–46.
6. Schlesinger, *A World Trimmed with Fur*, 53.
7. Frost, *Bering*, 90–91; Irish, *Hokkaido*, 57.
8. Frost, *Bering*, 177–78; Black, *Russians in Alaska*, 40–44.
9. Black, *Russians in Alaska*, 45–48.
10. Steller, *Journal*, 148.
11. Black, *Russians in Alaska*, 49. As Black notes, the exact number of sea otter skins that the survivors returned with is not known (*Russians in Alaska*, 56), although Steller claimed that they had "upward of 800" (*De bestiis marinis*).
12. Black, *Russians in Alaska*, 59–62.
13. McCracken, *Hunters of the Stormy Sea*, 32.
14. Vinkovetsky, *Russian America*, 31.
15. Black, *Russians in Alaska*, 66.
16. Black, *Russians in Alaska*, 70.
17. Jones, *Empire of Extinction*, 84.
18. Coxe, *Account*, 36.
19. Black, *Russians in Alaska*, 70.

20. Frost, *Bering*, 193–94; Jones, *Empire of Extinction*, 79. As Jones notes, Aleuts referred to a one-hatched canoe as an *igax* and the two-hatched variety as *ulyuxtax*; Russians collectively referred to them as *baidarki*. For a classic description of Aleut sea otter hunting, see Ogden, *The California Sea Otter Trade*, 11–14.
21. Coxe, *Account*, 12.
22. Shaw, *Musei Leveriani Explication*, 112.
23. Berkh, *A Chronological History*, 76.
24. Black, *Russians in Alaska*, 69.
25. Frost, *Bering*, 276–77.
26. Frost, *Bering*, 278; Urness, "Russian Mapping," 134.
27. Gibson, "The Exploration," 341–42.
28. Venegas, *A Natural and Civil History*, 222.
29. Cook, *Flood Tide of Empire*, 46.
30. Saunt, *West of the Revolution*, 56–58.
31. McDougall, *Let the Sea Make a Noise*, 64–65; Mapp, *The Elusive West*, 398–404.
32. Beebe and Senkewicz, *Lands of Promise and Despair*, 111.
33. For critical appraisals of the missions of Alta California, see Sandos, *Converting California*; Hackel, *Children of Coyote*.
34. Saunt, *West of the Revolution*, 72–74; Gibson, "The Exploration," 351; Chapman, *A History of California*, 272–74.
35. Engstrand, "Seekers," 93–94.
36. "Juan Perez's 'Diario,' 11 June–28 August 1774," in Beals, *Juan Perez*, 89.
37. Reid, *The Sea Is My Country*, 61–63.
38. Engstrand, "Seekers," 95.
39. Landgon, "Efforts at Humane Engagement."
40. Black, *Russians in Alaska*, 102–7.
41. Solovjova and Vovnyanko, "The Rise and Decline."
42. Black, *Russians in Alaska*, 107.
43. Matthews, *Glorious Misadventures*, 64–65.
44. Shelikhov, *A Voyage to America*, 45.
45. Miller, "Russian Routes."
46. Ogden, *The California Sea Otter Trade*, 2.
47. Ogden, *The California Sea Otter Trade*, 16–17.
48. Ogden, *The California Sea Otter Trade*, 18–24.
49. Ogden, *The California Sea Otter Trade*, 25.
50. Cook, *Flood Tide of Empire*, 353–55.
51. Gibson, "Nootka and Nutria," 158.
52. Cook, *Flood Tide of Empire*, 136.

53. Ogden, *The California Sea Otter Trade*, 24; Ogden, "The Californias," 459. Another example as to why Ogden's 1979 list needs to be used carefully is found in an eighteenth-century Spanish manuscript, reproduced by Warren Cook in 1973, that lists 3,356 skins taken to Asia in 1791 onboard the *Princess Real* (the majority of which were Vasadre's remaining pelts from California), a vessel that for some reason was not included by Ogden. See "Appendix C—A Balance Sheet of the Spanish Sea Otter Trade, 1786–1797," in Cook, *Flood Tide of Empire*, 549, see also 297–98.
54. Jones, *Empire of Extinction*, 153–54; Gibson, "The Exploration," 343. For the maritime effort known as the Mulkovsky expedition, see King, "The Mulkovsky Expedition."
55. Cook, *Flood Tide of Empire*, 111–15; Inglis, "The Effect of Laperouse," 48–49.
56. Miller, "The International Law of Discovery," 197; Urness, "Russian Mapping," 142.
57. Ogden, *California Sea Otter Trade*, 57–60. For Russian California, see Farris, *So Far from Home*; Schwartz, "Fort Ross, California."
58. Ogden, *California Sea Otter Trade*, 63–64. Diplomatic protest from Madrid to St. Petersburg regarding the existence of Fort Ross ultimately proved futile for imperial officials. See Taylor, "Spanish-Russian Rivalry," 124–27.
59. Cook *Flood Tide of Empire*, 532–33; Beebe and Senkewicz, *Lands of Promise*, 294.
60. Jones, *Empire of Extinction*, 61.
61. Jones, *Empire of Extinction*, 81.
62. Jones, *Empire of Extinction*, 82–83.
63. Dmytryshyn, Crownhart-Vaughan, and Vaughan, *The Russian American Colonies*, 146–48, 155. The settlement at Sitka was destroyed by Natives in 1802 but was rebuilt two years later. Yakutat was permanently devastated in 1805.
64. Ravalli, "Graphing the Sea Otter Hunt." Due to his Alaskan emphasis, Lensink did not include vessels that hunted in the Commander Islands in the early 1740s in his tally of ships, which begins at 1747 ("The History and Status," 10). For this and other reasons, he somewhat underestimated the environmental damage of eighteenth-century Russian otter hunting.
65. Jones, *Empire of Extinction*, 157.
66. Langsdorff, *Remarks and Observations*, 45.
67. Langsdorff, *Remarks and Observations*, 47.
68. Gibson, "Nootka and Nutria," 148, 153.
69. Jones, "Running into Whales," 363.
70. Dean, "The Sea Otter War," 25–28.
71. Jones, *Empire of Extinction*, 207–8.
72. Tikhmenev, *A History*, 405.

73. Jones, *Empire of Extinction*, 220.
74. Jones, *Empire of Extinction*, 210–13.
75. Jones, *Empire of Extinction*, 223–24.
76. For serial depletion and Russian environmental policies, see Bodkin, "Historic and Contemporary Status," 45–47. Somewhat confusingly, a separate essay coauthored by Bodkin claims that *no* effective harvest management took place during the maritime fur trade, which may indicate disagreement regarding the issue among contemporary biologists. See Larson and Bodkin, "The Conservation of Sea Otters," 7. For nineteenth-century critiques of Russian American Company marine mammal conservation, see Jones, *Empire of Extinction*, 210–12, 224–27.

3. BOSTON MEN

1. Ravalli, "Grounds of Our Claim"; Larson, "William Sturgis"; Loring, "Memoir of William Sturgis," 458.
2. For my purposes, the Pacific Northwest includes areas roughly encompassing the boundaries of modern-day southeastern Alaska, British Columbia, Washington, Oregon, Idaho, and Montana. I exclude the rest of Alaska and northern California. For geographic shifts in the maritime fur trade and their effect in prompting different conceptions of the Northwest, see Gibson, *Otter Skins*, 204–7; Mackie, *Trading beyond the Mountains*, 123–24.
3. See Furstenberg, "The Significance."
4. Heffer, *The United States and the Pacific*, 92.
5. Sturgis, "Examination."
6. Williams, "James Cook."
7. Beaglehole, *The Journals*, 296.
8. Matsuda, *Pacific Worlds*, 139–41; Gibson, "The Exploration," 380.
9. Lamb and Bartroli, "James Hanna," 3–6; Gibson, *Otter Skins*, 23.
10. Gibson, *Otter Skins*, 166. For conflicting information regarding the deadly encounter between the *Sea Otter* and the people of Nootka Sound, see Clayton, *Islands of Truth*, 69–70.
11. Dixon, *A Voyage round the World*, 235. For more on the sea otter trade of Haida Gwaii, see Sloan and Dick, *Sea Otters of Haida Gwaii*, 41–69.
12. Dixon, *A Voyage round the World*, 303.
13. Gibson, *Otter Skins*, 25–26.
14. Gibson, *Otter Skins*, 100.
15. Dolin, *When America First Met China*, 3–23, 72–87; Gibson and Whitehead, *Yankees in Paradise*, 95.
16. Gibson and Whitehead, *Yankees in Paradise*, 173–74.

17. Gibson, *Otter Skins*, 95.
18. Scofield, *Hail, Columbia*, 95–113.
19. Scofield, *Hail, Columbia*, 130–32.
20. Heffer, *The United States and the Pacific*, 29. Spanish explorer Bruno de Hezeta noticed the Columbia's offshore currents in 1775 and was the first non-Native to declare the existence of the river. Mears tried but failed to locate the same entrance in 1788. See Gough, *Fortune's a River*, 350–53.
21. Cook, *Flood Tide of Empire*, 143–44.
22. Colnett, *The Journal*, 20–21.
23. Scofield, *Hail, Columbia*, 147–49.
24. Tovell, *Voyage*, 34–36.
25. Cook, *Flood Tide of Empire*, 247.
26. Clayton, *Islands of Truth*, 178–79.
27. Clayton, *Islands of Truth*, 181.
28. Fichter, *So Great a Proffit*, 215.
29. Tovell, *Voyage*, 60–69; Cook, *Flood Tide of Empire*, 362–75. For the explorations of George Vancouver in the Pacific Northwest, see Fisher and Johnston, *From Maps to Metaphors*; Barnett, "The End."
30. Cook, *Flood Tide of Empire*, 412–13.
31. "Appendix I—The Northwest Fur Trade, by the Hon. William Sturgis," in Busch and Gough, *Fur Traders*, 93–94.
32. Gibson, *Otter Skins*, 214–16.
33. Gibson, *Otter Skins*, 121.
34. Gibson, *Otter Skins*, 128.
35. "Appendix II—William Sturgis on the American Vessels and the Maritime Trade," in Busch and Gough, *Fur Traders*, 107.
36. Gibson, *Otter Skins*, 58.
37. Gibson, *Otter Skins*, 30.
38. Mackie, *Trading beyond the Mountains*, 127.
39. Dolin, *When America First Met China*, 104–16; Fichter, *So Great a Proffit*, 218–20.
40. Dolin, *When America First Met China*, 123; Fichter, *So Great a Proffit*, 208–9.
41. See Matsuda, *Pacific Worlds*, 176–96.
42. Scofield, *Hail, Columbia*, 253–54.
43. Gibson, *Otter Skins*, 160–63; Igler, *The Great Ocean*, 77–79.
44. Gibson, *Otter Skins*, 171.
45. Jewitt, *A Narrative*.
46. Zilberstein, "Objects of Distant Exchange," 591–620.
47. Reid, *The Sea Is My Country*, 77–79.

48. "Appendix I—The Northwest Fur Trade, by the Hon. William Sturgis," in Busch and Gough, *Fur Traders*, 86.
49. Clayton, *Islands of Truth*, 156.
50. Reid, *The Sea Is My Country*, 55–56.
51. Dolin, *When America First Met China*, 130.
52. For a world systems approach to the sea otter trade, see Carlson, "The 'Otter Man' Empires," 390–442.
53. Phelps, "Solid Men," 59.
54. Gibson, *Otter Skins*, 44–45.
55. Gibson, *Otter Skins*, 48.
56. Igler, "Diseased Goods," 699–700.
57. Gibson, *Otter Skins*, 278–80.
58. Corney, *Early Voyages*, 87.
59. Corney, *Early Voyages*, 91–92.
60. Ogden, *The California Sea Otter Trade*, 45–47; Farris, "Otter Hunting"; Giesecke, "Unlikely Partners."
61. Ogden, *The California Sea Otter Trade*, 50–51.
62. Jones, "Running into Whales," 363–67.
63. Dmytryshyn, Crownhart-Vaughan, and Vaughan, *The Russian American Colonies*, 160. Such negative observations were made despite (or in fact because of) the fact that Americans were major suppliers of the Sitka colony at the time. For tensions arising from the sea otter trade as a reason for Russia dispatching the first official diplomat to the United States in 1808, see Kushner, *Conflict*, 11–12.
64. Block, "New England Merchants," 599–600. According to Block, it is difficult to determine exactly when Gale began in the hide and tallow trade. For the classic account of the trade, see Dana, *Two Years before the Mast*. For the influence of foreign immigration in Mexican California, see Nunis, "Alta California's Trojan Horse."
65. Busch and Gough, *Fur Traders*, 26. A similar attempt by a Russian crew to establish a base at the Columbia had failed just a few years before due to a shipwreck along the Olympic Peninsula. See Owens with Petrov, *Empire Maker*, 218–20.
66. Ronda, *Astoria and Empire*, 24–35.
67. Ronda, *Astoria and Empire*, 44.
68. Ronda, "The Education," 18; Fichter, *So Great a Proffit*, 272–75.
69. Mackie, *Trading beyond the Mountains*, 14.
70. Clayton, *Islands of Truth*, 206–8.
71. Ronda, *Astoria and Empire*, 327–35; Graebner, *Empire on the Pacific*, 23–24.
72. Kushner, *Conflict*, 15.

73. Kushner, *Conflict*, 17–19; Ronda, *Astoria and Empire*, 71–73.
74. Kushner, *Conflict*, 22. Furthermore, as Sturgis argued in his 1822 response to the *ukase* (discussed below), such arrangements undercut the Russian argument that trading by American vessels was "illicit" ("Examination," 394).
75. Kushner, *Conflict*, 26; Gibson, *Otter Skins*, 263.
76. Kushner, *Conflict*, 30. Floyd's connection to western expansion included his longtime friendship with explorer William Clark. Benton's Missouri connections made him a strong supporter of the overland fur trade, and he developed a close relationship with Astor.
77. Benton, *Thirty Years View*, 13. For more on the effect of Floyd's proposed bill within American political circles, see Kushner, *Conflict*, 30, 41; Ronda, *Astoria and Empire*, 331–33.
78. Black, *Russians in Alaska*, 191–92.
79. For a sampling of scholarship on the question, see Pletcher, *The Diplomacy of Involvement*, 20–21n23.
80. Kushner, *Conflict*, 34.
81. For Adams's relationships with Boston sea otter merchants, see Barragy, "American Maritime Otter Diplomacy," chap. 2. For early American whalers in the Pacific, see Gibson and Whitehead, *Yankees in Paradise*, chap. 8; Heffer, *The United States and the Pacific*, chap. 2.
82. Malloy, "A Most Remarkable Enterprise," 107–15.
83. Secretary of State Adams wrote to the American minister in Russia, Henry Middleton, in early 1823 and instructed him to read Sturgis's article in order to strengthen his stance in St. Petersburg. The Russian consul in the United States sent two copies of it to the Russian foreign minister (Kushner, *Conflict*, 171–72n37). In addition, the publication *Niles Weekly Register* reprinted portions of the article in November 1822 in criticism of the *ukase*, and Senator James Lloyd of Massachusetts wrote to President James Monroe in May 1823, citing Sturgis for the importance of Pacific commerce to the nation (Kushner, *Conflict*, 40–41, 54).
84. Sturgis, "Examination," 380–83.
85. Sturgis, "Examination," 390.
86. Kushner, *Conflict*, 59–60.
87. Black, *Russians in Alaska*, 198.
88. Mackie, *Trading beyond the Mountains*, 30–31.
89. Gibson, *Otter Skins*, 62.
90. For details on the British fur trade in the Pacific at this time, see Mackie, *Trading beyond the Mountains*, chaps. 3, 5, and 6. For the Hudson's Bay Company's lease in southeastern Alaska, which allowed the company to replace American

vessels as suppliers for Russian America, see Gibson, *Otter Skins*, 78–80; Jackson, "The Stikine Territory Lease."
91. Simpson, *An Overland Journey*, 131.
92. Mackie, *Trading beyond the Mountains*, 142–45. For the company's Northern Department environmental policies, see Ray, "Some Conservation Schemes."
93. Mackie, *Trading beyond the Mountains*, 144–45; Ogden, *Otter Skins*, 128–30.
94. Sowards, *United States West Coast*, 223–25.
95. Ravalli, "Grounds of Our Claim," 20–22.
96. *Congressional Globe*, 28th Cong., 2nd sess., 226. For similar statements emphasizing the value of Pacific trade, see Graebner, *Empire on the Pacific*, 37–38.
97. Ogden, *The California Sea Otter Trade*, 141–42.
98. Ogden, *The California Sea Otter Trade*, 91–93, 116; Woolfenden and Elkinton, *Cooper*.
99. Ogden, *The California Sea Otter Trade*, 113.
100. Job Francis Dye, "Recollections of California," 1–2, Bancroft Library, University of California, Berkeley. In another document, Pardo is spelled "Pecao," and Dye recalled collecting "about twenty-four otter skins" during his brief partnership with him. See Job Dye, "Early History and Reminiscence of Job Francis Dye," Holt-Atherton Special Collections, University of the Pacific, Stockton, California.
101. Dye, "Recollections," 2–3; Ogden, *The California Sea Otter Trade*, 111.
102. Jones, *Empire of Extinction*, 211.
103. There is some evidence that sea otter pups may have been less frequently targeted during Mexican California than during the Spanish era. A catalog of California otter trade vessel cargoes (see the appendix) suggests that furs of baby otters were exported more often before the 1820s than after.
104. Beidleman, *California's Frontier Naturalists*, 309–11; Audubon and Bachman, *The Quadrupeds*.
105. Audubon and Bachman, *The Quadrupeds*, 174.

4. NEAR EXTINCTION AND REEMERGENCE

1. Buxton, "The Sea Otter Hunters," 13–14; Stephan, *The Kuril Islands*, 99; Murray, *The Vagabond Fleet*, 21–22; Kimberly, "50 Years and More"; Jordan, *Seal and Salmon Fisheries*, 74–78, 88–90; "Kimberly, Martin Morse."
2. VanBlaricom, "Synopsis."
3. Keiner, "How Scientific."
4. Gibson and Whitehead, *Yankees in Paradise*, 181.
5. For an accessible account of the Opium War, see Dolin, *When America First Met China*, 209–64.

6. Gibson and Whitehead, *Yankees in Paradise*, 183–84; Dolin, *When America First Met China*, 285.
7. Ogden, *The California Sea Otter Trade*, 148–50.
8. Black, *Russians in Alaska*, 282. Increased Russian textile exports to Kyakhta and the opening of the port of Shanghai led to the decline of the Kyakhta fur trade by 1850. See Gibson, "Sitka–Kyakhta versus Sitka–Canton," 78–79.
9. "The Costly Sea Otter," *New York Times*, October 29, 1894, 10; "Furs Are Dearer," *London Daily Mail*, May 28, 1899, 6.
10. Scofield, *Hail, Columbia*, 348.
11. Scammon, *The Marine Mammals*, 174.
12. Scheffer, "The Last of the Sea Otter Hunters," 15.
13. Scheffer, "The Last of the Sea Otter Hunters," 15–16.
14. Scheffer, "The Sea Otter," 372–73.
15. Kenyon, *The Sea Otter*, 184–85; Hatch, "Elakha," 82–84; "Kill Three Sea Otter near the North Jetty," *Aberdeen Herald*, June 20, 1913, 5. According to Kenyon, the obscurity of the Oregon information suggests that sea otters may have gone extinct there as early as the late nineteenth century. As the *Aberdeen Herald* report shows, Kenyon's date of 1910 for the killing of the last Washington sea otter is perhaps too early.
16. Kenyon, *The Sea Otter*, 183. Once again, Kenyon was apparently unaware of a kill at Vancouver Island in 1931, which is probably the last of the species verified to have been taken by a hunter in coastal British Columbia. See Nichol, "Conservation in Practice," 375.
17. Gallo-Reynoso, "Status of Sea Otters."
18. Black, *Russians in Alaska*, 273–75. For an overview of Russian historiography of motivations for the sale of Alaska emphasizing the economic instability of the Russian American Company, see Grinev, "Why Russia Sold Alaska." For a more recent study of the purchase, see Farrow, *Seward's Folly*.
19. Lee, "Context and Contact," 23–27.
20. Lee, "Context and Contact," 28.
21. Lee, "Context and Contact," 30–35.
22. Kahn, *Jewish Voices*, 278–81; Muller, "Appendix V," 614.
23. Lee, "Context and Contact," 33.
24. VanBlaricom, "Synopsis," 397–98; *Appendix to the Congressional Globe*, 41st Cong., 2nd sess., 675.
25. Wright, "The Sea Otter Industry," 257–58; VanBlaricom, "Synopsis," 398.
26. Hooper, *A Report*, 10–11.

27. Elliott, *Our Arctic Province*, 127–44; Wright, "The Sea Otter Industry," 259. Elliott described shooting at sea otters from the shore as "heterodoxy . . . [that] has only been in vogue for a short time" (*Our Arctic Province*, 141). Yet he also realized the long-term effect it would have in the "virtual extermination" of the animals in Alaska (*Our Arctic Province*, 130). Hooper surmised that guns in Alaska, "while not more destructive than the spear," were perhaps more wasteful weapons that produced injured animals that eluded capture and later died (*A Report*, 8).
28. Hooper, *A Report*, 16.
29. Elliott, "The Sea-Otter Fishery," 486.
30. Wright, "The Sea Otter Industry," 59; Jordan, "Colonial Lessons of Alaska," 581. For Aleut complaints against stationary white sea otter hunters, see Black, "The Nature of Evil," 138–39.
31. Busch, *The War against the Seals*, 135–38. Conversely, ships designated as pelagic sealers occasionally pursued sea otters. See Murray, *The Vagabond Fleet*, 209.
32. Hooper, *A Report*, 12.
33. Wright, "The Sea Otter Industry," 258–62.
34. Tikhmenev, *A History*, 431.
35. Emmons, *The Tlingit Indians*, 123.
36. Lensink, "The History and Status," 17.
37. Stephan, *The Kuril Islands*, 99; Stone, "Hunting Marine Mammals," 47.
38. Stone, "Hunting Marine Mammals," 49–50; Stephan, *The Kuril Islands*, 99–103.
39. Snow, *In Forbidden Seas*, 296.
40. "Hunting the Sea Otters," *Pacific States Watchman*, October 15, 1881, 335, California History Room, California State Library, Sacramento.
41. Snow, *In Forbidden Seas*, 81.
42. McCracken, *Hunters of the Stormy Sea*, 287–88.
43. Nichol, "Conservation in Practice," 375.
44. Lindsay, "The Passing of the Sea Otter," *Arizona Republican*, October 24, 1907, 9. Conversely, netting was believed by Hooper to be an inefficient hunting practice when utilized by white hunters in Alaska because animals could become trapped in nets for days and attacked by crustaceans before being collected, thus ruining their pelts (*A Report*, 8).
45. Lavrov, "Fur Treasures," 74; Marakov, "A Natural Monument." For early Soviet nature protection, see Weiner, *A Little Corner of Freedom*, chaps. 1 and 2.
46. Dasman, "Environmental Changes," 110.
47. Rogers, "Memoirs," 47–50; Elliott, "The Sea-Otter Fishery," 487; Buxton, "The Sea Otter Hunters," 17.
48. See Taylor, "The Mining Boom," 463–92.
49. Ogden, *The California Sea Otter Trade*, 139.

50. Snow, *In Forbidden Seas*, 292–93.
51. Snow, *In Forbidden Seas*, 52–53.
52. Hinckley, *The Americanization of Alaska*, 88–94.
53. For the Santa Cruz Island Foundation data, see "Sea Otters." For sea otters as "completely eradicated" from the Channel Islands by the middle of the nineteenth century, see Braje, *Shellfish for the Celestial Empire*, 100.
54. "A Misconception," *Daily Independent*, June 22, 1885. For Chinese fisheries in California, see Armentrout-Ma, "Chinese in California's Fishing Industry."
55. "Furs Are Dearer," *London Daily Mail*, May 28, 1899, 6.
56. Goetzmann and Sloan, *Looking Far North*, 3–15.
57. Goetzmann and Sloan, *Looking Far North*, 201–6.
58. Frost, "Crosscurrents and Deep Water," 173.
59. Frost, "Crosscurrents and Deep Water," 174. A website for the volume in which Frost's essay appears claims that nobody in 1899 saw a sea otter, but this may be an exaggeration. See "Sea Otters in Alaska."
60. Frost, "Crosscurrents and Deep Water," 173.
61. Dorsey, *The Dawn of Conservation Diplomacy*, 111–15.
62. Dorsey, *The Dawn of Conservation Diplomacy*, 119–24.
63. For the disputes between Elliott and David Starr Jordan, see Dorsey, *The Dawn of Conservation Diplomacy*, chap. 5.
64. Jordan, *Seal and Salmon Fisheries*, 60.
65. Busch, *The War against the Seals*, 123–26.
66. Busch, *The War against the Seals*, 126.
67. Dorsey, *The Dawn of Conservation Diplomacy*, 159.
68. "Convention between the United States."
69. VanBlaricom, "Synopsis," 428.
70. VanBlaricom, "Synopsis," 429.
71. Elliott, "The Sea-Otter Fishery," 483–84.
72. VanBlaricom, "Synopsis," 409–10; Nichol, "Conservation in Practice," 374–75. The continuation of Japanese commercial harvests prior to World War II may have slowed sea otter recovery in the Kurils, although this requires more research. Reportedly, there were more than seven hundred animals there by the 1930s. See Nichol, "Conservation in Practice," 374; Kornev and Korneva, "Population Dynamics," 275; Kornev and Korneva, "Historical Trends," 21.
73. Murie, "Notes," 119–31; Glover, "Sweet Days," 138; Kenyon, *The Sea Otter*, 1.
74. VanBlaricom, Belting, and Triggs, "Sea Otters," 204–5.
75. VanBlaricom, Belting, and Triggs, "Sea Otters," 205–7.
76. Bryant, "Sea Otters," 135.
77. Oyer, "Sea Otters," 88; Farnsworth, "Sea Otters," 90.

78. See Leatherwood, Harrington-Coulombe, and Hubbs, "Relict Survival," 110.
79. Sharpe, "The Discovery."
80. Bolin, "Reappearance," 301; "Sea Otter Herd Is Sighted," *Monterey Peninsula Herald*, March 28, 1938, Sea Otters Clippings Files, California Room, Monterey Public Library, Monterey, California.
81. For Sharpe's connection to the coastal construction, see Newland, "Bixby Creek Bridge."
82. Macdonald, *Pacific Pelts*.
83. "Sea Otter Herd Is Sighted."
84. Palumbi and Sotka, *The Death and Life*, 117–20, 128–29.
85. Estes et al., "Causes of Mortality," 198–216.
86. Kenyon, *The Sea Otter*, 282.
87. Fisher, "Prices of Sea Otter Pelts," 264.
88. Estes et al., "Causes of Mortality," 206.
89. Palumbi and Sotka, *The Death and Life*, 115.
90. Estes et al., "Causes of Mortality," 212.
91. Estes et al., "Causes of Mortality," 212–13.

5. NUKES, AQUARIA, AND CUTENESS

1. *Otters Holding Hands*; "Vancouver Sea Otters a Hit on YouTube," CBC *News*, April 3, 2007, http://www.cbc.ca/news/canada/british-columbia/vancouver-sea-otters-a-hit-on-youtube-1.688725; Del Bucchia, *Coping with Emotions*.
2. VanBlaricom, *Sea Otters*, 8.
3. Kohlhoff, *Amchitka and the Bomb*, 16–17.
4. Kohlhoff, *Amchitka and the Bomb*, 31–32; Jones, "Present Status," 376–83.
5. Brinkley, *The Quiet World*, 427–30.
6. Kohlhoff, *Amchitka and the Bomb*, 38; Kinney, "The Otters of Amchitka," 295.
7. Jones, "Present Status," 378.
8. Murphy, "Tales of the Sea Otter," 30, 128–30.
9. "The Playful Sea Otter," *Outdoor California*, August 1958, 12.
10. Murphy, "Tales of the Sea Otter," 30. Other periodicals in the first half of the twentieth century such as *National Geographic* continued to highlight the commercial value of marine mammals. See Lavigne, Scheffer, and Kellert, "The Evolution," 24.
11. Kinney, "Otters of Amchitka," 300–301; Weyler, *Greenpeace*, 137.
12. Kinney, "Otters of Amchitka," 293.
13. "Uproar over Otters," *Life*, October 15, 1965, 151–52.
14. "AEC to Test Nuclear Warhead in Wildlife Refuge," *Sierra Club Bulletin*, July 1969, 3. As Weyler notes, national Sierra Club leaders in San Francisco failed

to endorse the activities of the club's Vancouver branch members that resulted in the formation of the Don't Make a Wave Committee, citing nuclear disarmament as a "distraction" from environmental advocacy (*Greenpeace*, 69).

15. Kohlhoff, *Amchitka and the Bomb*, 110; Ballachey and Bodkin, "Challenges," 71.
16. Kohlhoff, *Amchitka and the Bomb*, 38; 46–47; Bodkin, "Historic and Contemporary Status," 50–51. Nevertheless, as noted below, the efforts of Jones and others in the 1950s helped to bring some of the first sea otters (from Amchitka) to be held in captivity at American facilities.
17. Bodkin, "Historic and Contemporary Status," 50; Nichol, "Conservation in Practice," 376–79.
18. Bodkin, "Historic and Contemporary Status," 51–52; Rauzon, "Memories," 998; Kenyon, *The Sea Otter*.
19. See Estes, *Serendipity*, chaps. 4 and 6.
20. NNSANevada, *The Warm Coat*.
21. Kohlhoff, *Amchitka and the Bomb*, 91–92.
22. Stephan, *The Kuril Islands*, 173–75.
23. Nichol, "Conservation in Practice," 374; Kornev and Korneva, "Population Dynamics," 275; Kornev and Korneva, "Historical Trends," 21.
24. Stephan, *The Kuril Islands*, 177.
25. Ivashchenko and Clapham, "Soviet Illegal Whaling." Records made available in the 1990s confirmed that Soviets reported fraudulent whale catches to international regulators. See Dorsey, *Whales and Nations*, 156.
26. Peebles, *The Undersea World*; *Mutual of Omaha's Wild Kingdom*.
27. "We All Love Sea Otters," *Independent Coast Observer*, April 1, 1970, http://ico.stparchive.com/Archive/ICO/ICO04011970p01.php.
28. Margaret Owings, "Do Sea Otters Have Any Friends?," Sea Otters Clippings Files, California Room, Monterey Public Library, Monterey, California; Carswell, Speckman, and Gill, "Shellfish Fishery Conflicts," 340.
29. John Woolfenden, "The Darlings of Tourists," *Herald Weekend Magazine*, October 21, 1972, 10, Sea Otters Clippings Files, California Room, Monterey Public Library, Monterey, California.
30. Gilbert, "Dept. of Otter Confusion," 72.
31. For a summary of sea otter management issues in California following the expansion of federal environmental policy, see Carswell, Speckman, and Gill, "Shellfish Fishery Conflicts," 343–49.
32. Baur, Bean, and Gosliner, "The Laws," 49; VanBlaricom, "Synopsis," 405.
33. VanBlaricom, "Synopsis," 405–7. For background on the Marine Mammal Protection Act, see Ray and Potter, "The Making," 522–52. For the Endangered Species Act, see Alagona, *After the Grizzly*, chap. 4.

34. Carswell, Speckman, and Gill, "Shellfish Fishery Conflicts," 353.
35. Ballachey and Bodkin, "Challenges," 68.
36. Ross Perlin, "Why Would Anyone Want to Shoot a Sea Otter?," *Guardian*, March 10, 2015, https://www.theguardian.com/world/2015/mar/10/why-would-anyone-want-to-shoot-a-sea-otter. Also see Vanessa Friedman, "Is All Fur Bad?," *New York Times*, December 1, 2016, https://www.nytimes.com/2016/12/01/fashion/sea-otter-fur-hunting-alaska-fashion-debate.html?_r=0.
37. Nichol, "Conservation in Practice," 379.
38. Salomon et al., "First Nations Perspectives," 306–7; Nichol, "Conservation in Practice," 379.
39. On southeastern Alaska, see Carswell, Speckman, and Gill, "Shellfish Fishery Conflicts," 350–62.
40. VanBlaricom, Belting, and Triggs, "Sea Otters in Captivity," 199–200; Kirkpatrick, Stullken, and Jones, "Notes," 47.
41. Vincenzi, "The Sea Otter," 28.
42. Chiang, *Shaping the Shoreline*, 155–57.
43. Chiang, *Shaping the Shoreline*, 170.
44. Chaing, *Shaping the Shoreline*, 173.
45. Mary Rodriguez, "Significant Otters," *Monterey Bay Aquarium Newsletter*, September 1987, Sea Otters Clippings Files, California Room, Monterey Public Library, Monterey, California.
46. Kawata, "Zoological Gardens," 315; Yoshinori Yasui, "Will Sea Otters Disappear from Japanese Aquariums?," *AsiaOne*, April 28, 2014, http://news.asiaone.com/news/asia/will-sea-otters-disappear-japanese-aquariums.
47. Yasui, "Will Sea Otters Disappear"; "Hard-to-Breed Sea Otters Set to Vanish from Nation's Aquariums," *Japan Times*, April 3, 2016, http://www.japantimes.co.jp/news/2016/04/03/national/hard-breed-sea-otters-set-vanish-nations-aquariums/#.WKE7nDsrKM9.
48. Ballachey and Bodkin, "Challenges," 80. For more on the *Exxon Valdez*, see Day, *Red Light*.
49. Batten, "Press Interest," 33.
50. Batten, "Press Interest," 35.
51. Ballachey and Bodkin, "Challenges," 72–73.
52. Ballachey and Bodkin, "Challenges," 73–74. As noted in the previous chapter, the Bixby Creek sea otters of 1938 were reportedly attacked by a killer whale, suggesting that orca activity may have been a reason for the relative isolation of the central California herd.
53. Estes, *Serendipity*, 188.

54. Nickerson, *Sea Otters*, 13, 26.
55. VanBlaricom, *Sea Otters*, 26.
56. Leon, *A Raft of Sea Otters*, 5.
57. Murray, *Sea Otters*, 11.
58. Silverstein, Silverstein, and Silverstein, *The Sea Otter*, 7.
59. Leila Kheiry, "Stedman Talks Oil Tax, Sea Otters," *KRBD*, April 17, 2013, http://www.krbd.org/2013/04/17/stedman-talks-oil-tax-sea-otters.
60. Naomi Klouda, "A Bounty on Sea Otters?," *Homer Tribune*, March 30, 2013, http://homertribune.com/2013/03/a-bounty-on-sea-otters/.
61. Stevens, Organ, and Serfass, "Otters as Flagships."
62. David Perlman, "Cuddly Otter Image in the Tank," *San Francisco Chronicle*, February 4, 1995, A20.
63. Perlman, "Cuddly Otter Image."
64. Brian Switek, "Sea Otters Are Jerks," *Slate*, October 28, 2013, http://www.slate.com/blogs/wild_things/2013/10/28/sea_otter_dolphin_and_penguin_behavior_your_favorite_animals_are_jerks.html.
65. Larry Pynn, "Fifty Shades of Fur? Exposing the Dark Side of a Sea Otter's Sex Life," *Vancouver Sun*, April 10, 2014, http://www.vancouversun.com/technology/Fifty+shades+Exposing+dark+side+otter+life+with+video/9721555/story.html.
66. For an introduction to *kawaii* culture, see Sato, "From Hello Kitty," 38–42.
67. Cynthia Holmes, interview with the author, May 3, 2017.
68. Del Bucchia, *Coping with Emotions*, 93.
69. Del Bucchia, *Coping with Emotions*, 95.

CONCLUSION

1. Catton, *Land Reborn*; Williams and Hooten, "The Extraordinary Return."

APPENDIX

1. Ogden, *The California Sea Otter Trade*; Ogden, "Trading Vessels on the California Coast, 1786–1848," Bancroft Library, University of California, Berkeley.
2. Farris, "Otter Hunting," 20–33.
3. A version of this appendix was originally published by Kirsten Livingston, and Hannah Zimmerman and the author as a research note in *International Journal of Maritime History* in 2012. A number of corrections were subsequently made and are offered here. Any discrepancies that remain between Ogden's data and this appendix are the author's fault alone.
4. Jones et al., "Toward a Prehistory," 243–71.

BIBLIOGRAPHY

Abé, Takao. "The Seventeenth Century Jesuit Missionary Reports on Hokkaido." *Journal of Asian History* 39, no. 2 (2005): 111–28.
Alagona, Peter S. *After the Grizzly: Endangered Species and the Politics of Place in California.* Berkeley: University of California Press, 2013.
Appendix to the Congressional Globe. 40th Cong., 2nd sess., 1868.
Appendix to the Congressional Globe. 41st Cong., 2nd sess., 1870.
Armentrout-Ma, L. Eve. "Chinese in California's Fishing Industry, 1850–1914." *California History* 60 (1981): 142–57.
Armitage, David, and Alison Bashford, eds. *Pacific Histories: Ocean, Land, People.* Basingstoke: Palgrave Macmillan, 2014.
Arndt, Katherine L. "Preserving the Future Hunt: The Russian-American Company and Marine Mammal Conservation Policies." *Fort Ross–Salt Point Newsletter,* Fall 2007, 4–6.
Audubon, John James, and John Bachman. *The Quadrupeds of North America.* Vol. 3. New York: George R. Lockwood, 1854.
Ballachey, Brenda E., and James L. Bodkin. "Challenges to Sea Otter Recovery and Conservation." In *Sea Otter Conservation,* edited by Shawn E. Larson, James L. Bodkin, and Glenn R. VanBlaricom. London: Elsevier, 2015.
Barnett, James K. "The End of the Northern Mystery: George Vancouver's Survey of the Northwest Coast." In *Arctic Ambition: Captain Cook and the Northwest Passage,* edited by James K. Barnett and David L. Nicandri, 263–87. Anchorage: Anchorage Museum, 2015.
Barragy, Terrence J. "American Maritime Otter Diplomacy." Ph.D. dissertation, University of Wisconsin, 1974.
Barthélemy de Lesseps, Jean-Baptiste. *Travels in Kamtschatka.* New York: Arno Press and the *New York Times,* 1970.
Batten, B. T. "Press Interest in Sea Otters Affected by the T/V *Exxon Valdez* Oil Spill." In *Sea Otter Symposium: Proceedings of a Symposium to Evaluate the Response*

Effort on Behalf of Sea Otters after the T/V Exxon Valdez Oil Spill into Prince William Sound, Anchorage, Alaska, 17–19 April 1990, edited by K. Bayha and J. Kormendy, 32–40. Washington DC: U.S. Department of the Interior, Fish and Wildlife Service, and National Fish and Wildlife Foundation, 1990.

Bauer, K. Jack. "Pacific Coastal Commerce in the American Period." *Journal of the West* 20, no. 3 (1981): 11–20.

Baur, Donald C., Michael J. Bean, and Michael L. Gosliner. "The Laws Governing Marine Mammal Conservation in the United States." In *Conservation and Management of Marine Mammals*, edited by John R. Twiss and Randall R. Reeves, 48–86. Washington DC: Smithsonian Institution Press, 1999.

Beaglehole, J. C., ed. *The Journals of Captain James Cook on His Voyages of Discovery*. Vol. 3. Cambridge: Hakluyt Society, 1967.

Beals, Herbert K, trans. and annotation. *Juan Perez on the Northwest Coast: Six Documents of His Expedition in 1774*. Portland: Oregon Historical Society Press, 1989.

Beebe, Rose Marie, and Robert M. Senkewicz, eds. *Lands of Promise and Despair: Chronicles of Early California, 1535–1846*. Berkeley: Heyday Books, 2001.

Beidleman, Richard G. *California's Frontier Naturalists*. Berkeley: University of California Press, 2006.

Benton, Thomas Hart. *Thirty Years View; or, A History of the Working of the American Government for Thirty Years, from 1820 to 1850*. Vol. 1. New York: D. Appleton and Company, 1858.

Berkh, Vasilii Nikolaevich. *A Chronological History of the Discovery of the Aleutian Islands*. Translated by Dmitri Krenov, edited by Richard A. Pierce. Kingston ON: Limestone Press, 1974.

Berta, Annalisa, James L. Sumich, and Kit M. Kovacs. *Marine Mammals: Evolutionary Biology*. 3rd ed. London: Elsevier, 2015.

Black, Lydia T. "Animal World of the Aleuts." *Arctic Anthropology* 35, no. 2 (1998): 126–35.

———. "The Nature of Evil: Of Whales and Sea Otters." In *Indians, Animals, and the Fur Trade: A Critique of Keepers of the Game*, edited by Shepard Krech III, 109–47. Athens: University of Georgia Press, 1981.

———. *Russians in Alaska, 1732–1867*. Fairbanks: University of Alaska Press, 2004.

Block, Michael D. "New England Merchants, the China Trade, and the Origins of California." Ph.D. dissertation, University of Southern California, 2011.

Bodkin, James L. "Historic and Contemporary Status of Sea Otters in the North Pacific." In *Sea Otter Conservation*, edited by Shawn E. Larson, James L. Bodkin, and Glenn R. VanBlaricom, 44–61. London: Elsevier, 2015.

Boessenecker, Robert W. "A Middle Pleistocene Sea Otter from North California and the Antiquity of *Enhydra* in the Pacific Basin." *Journal of Mammalian Evolution*, December 27, 2016, https://link.springer.com/article/10.1007/s10914-016-9373-6.

Bolin, Rolf L. "Reappearance of the Southern Sea Otter along the California Coast." *Journal of Mammalogy* 19, no. 3 (1938): 301.

Braje, Todd J. *Shellfish for the Celestial Empire: The Rise and Fall of Commercial Abalone Fishing in California.* Salt Lake City: University of Utah Press, 2016.

Brinkley, Douglas. *The Quiet World: Saving Alaska's Wilderness Kingdom, 1879–1960.* New York: HarperCollins, 2011.

Bryant, H. C. "Sea Otters near Point Sur." *California Fish and Game* 1, no. 2 (1915): 135.

Busch, Briton Cooper. *The War against the Seals: A History of the North American Seal Fishery.* Kingston, ON: McGill-Queen's University Press, 1985.

Busch, Briton C., and Barry M. Gough, eds. *Fur Traders from New England: The Boston Men in the North Pacific, 1787–1800.* Spokane WA: Arthur H. Clark Company, 1997.

Buxton, Michael. "The Sea Otter Hunters of San Diego and the Lower Coast, 1846–1903." *Mains'l Haul: A Journal of Pacific Maritime History* 43, nos. 3 and 4 (Summer/Fall 2007): 8–19.

Carlson, John D. "The 'Otter Man' Empires: The Pacific Fur Trade, Incorporation and the Zone of Ignorance." *Journal of World Systems Research* 3, no. 3 (Fall 2002): 390–442.

Carswell, Lilian P., Suzann G. Speckman, and Verena A. Gill. "Shellfish Fishery Conflicts and Perceptions of Sea Otters in California and Alaska." In *Sea Otter Conservation*, edited by Shawn E. Larson, James L. Bodkin, and Glenn R. VanBlaricom, 333–68. London: Elsevier, 2015.

Catton, Theodore. *Land Reborn: A History of Administration and Visitor Use in Glacier Bay National Park and Preserve.* Anchorage: National Park Service, 1995, https://www.nps.gov/parkhistory/online_books/glba/adhi/contents.htm.

Chanin, Paul. *The Natural History of Otters.* New York: Facts on File, 1985.

Chapman, Charles E. *A History of California: The Spanish Period.* New York: Macmillan Company, 1921.

Chiang, Connie Y. *Shaping the Shoreline: Fisheries and Tourism on the Monterey Coast.* Seattle: University of Washington Press, 2008.

Clayton, Daniel. *Islands of Truth: The Imperial Fashioning of Vancouver Island.* Vancouver: University of British Columbia Press, 2000.

Cochrane, John Dundas. *Narrative of a Pedestrian Journey through Russia and Siberian Tartary.* Vol. 2. New York: Arno Press and the *New York Times*, 1970.

Colnett, James. *The Journal of Captain James Colnett aboard the* Argonaut. Edited by F. W. Howay. Toronto: Champlain Society, 1940.

Congressional Globe, 28th Cong., 2nd sess., 1845.

"Convention between the United States, Great Britain, Russia and Japan for the Preservation and Protection of Fur Seals." *NOAA.org*, http://pribilof.noaa.gov/documents/THE_FUR_SEAL_TREATY_OF_1911.pdf.

Cook, Warren L. *Flood Tide of Empire: Spain and the Pacific Northwest, 1543–1819.* New Haven, CT: Yale University Press, 1973.

Corbett, Debra G., Douglas Causey, Mark Clementz, Paul L. Koch, Angela Doroff, Christine Lefevre, and Dixie West. "Aleut Hunters, Sea Otters, and Sea Cows: Three Thousand Years of Interactions in the Western Aleutian Islands, Alaska." In *Human Impacts on Ancient Marine Ecosystems: A Global Perspective,* edited by Torben C. Rick and Jon M. Erlandson, 43–75. Berkeley: University of California Press, 2008.

Corney, Peter. *Early Voyages in the North Pacific, 1813–1818.* Fairfield WA: Ye Galleon Press, 1965.

Coxe, William. *Account of the Russian Discoveries between Asia and America.* New York: Argonaut Press, 1966.

Dana, William Henry. *Two Years before the Mast.* New York: Signet Classic, 1964.

Dasman, Raymond. "Environmental Changes before and after the Gold Rush." *California History* 77, no. 4 (1998–99): 105–22.

Day, Angela. *Red Light to Starboard: Recalling the* Exxon Valdez *Disaster.* Pullman: Washington State University Press, 2014.

Dean, Jonathan R. "The Sea Otter War of 1810: Russia Encounters the Tsimshians." *Alaska History* 12, no. 2 (Fall 1997): 24–31.

Del Bucchia, Dina. *Coping with Emotions and Otters.* Vancouver BC: Talonbooks, 2013.

Dixon, George. *A Voyage round the World; but More Particularly to the North-west Coast of America: Performed in 1785, 1786, 1787, and 1788 . . .* London: Geo. Goulding, 1789.

Dmytryshyn, Basil, E. A. P. Crownhart-Vaughan, and Thomas Vaughan, eds. and trans. *The Russian American Colonies, 1798–1867: A Documentary Record.* Portland: Oregon Historical Society Press, 1989.

———. *Russian Penetration of the North Pacific Ocean, 1700–1799: A Documentary Record.* Portland: Oregon Historical Society Press, 1988.

Dolin, Eric Jay. *When America First Met China: An Exotic History of Tea, Drugs, and Money in the Age of Sail.* New York: Liveright Publishing, 2012.

Dorsey, Kurkpatrick. *The Dawn of Conservation Diplomacy: U.S.-Canadian Wildlife Protection Treaties in the Progressive Era.* Seattle: University of Washington Press, 1998.

———. *Whales and Nations: Environmental Diplomacy on the High Seas.* Seattle: University of Washington Press, 2013.

Elliott, Henry W. *Our Arctic Province: Alaska and the Seal Islands.* New York: Charles Scribner's Sons, 1886.

———. "The Sea-Otter Fishery." In *The Fisheries and Fishery Industries of the United States.* Vol. 2, sec. 5. Edited by George Brown Goode. Washington DC: Government Printing Office, 1887.

Emmons, George Thornton. *The Tlingit Indians*. Edited and with additions by Frederica de Laguna. Seattle: University of Washington Press, 1991.

Engstrand, Iris H. W. "Seekers of the 'Northern Mystery': European Exploration of California and the Pacific." In *Contested Eden: California before the Gold Rush*, edited by Ramon A. Gutierrez and Richard J. Orsi, 78–110. Berkeley: University of California Press, 1998.

Erlandson, Jon M., Madonna L. Moss, and Matthew Des Lauriers. "Life on the Edge: Maritime Cultures of the Pacific Coast of North America." *Quaternary Science Reviews* 27 (2008): 2236.

Estes, James A. "Natural History, Ecology, and the Conservation and Management of Sea Otters." In *Sea Otter Conservation*, edited by Shawn E. Larson, James L. Bodkin, and Glenn R. VanBlaricom, 19–41. London: Elsevier, 2015.

———. *Serendipity: An Ecologist's Quest to Understand Nature*. Oakland: University of California Press, 2016.

Estes, James A., Brian B. Hatfield, Katherine Ralls, and Jack Ames. "Causes of Mortality in California Sea Otters during Periods of Population Growth and Decline." *Marine Mammal Science* 19, no. 1 (2003): 198–216.

Farnsworth, G. "Sea Otters near Catalina Island." *California Fish and Game* 3, no. 2 (1917): 90.

Farris, Glenn. "Otter Hunting by Alaskan Natives along the California Coast in the Early Nineteenth Century." *Mains'l Haul: A Journal of Pacific Maritime History* 43, nos. 3 and 4 (Summer/Fall 2007): 20–33.

———. *So Far from Home: Russians in Early California*. Berkeley: Heyday Books, 2012.

Farrow, Lee. *Seward's Folly: A New Look at the Alaska Purchase*. Fairbanks: University of Alaska Press, 2016.

Fichter, James R. *So Great a Proffit: How the East Indies Trade Transformed Anglo-American Capitalism*. Cambridge MA: Harvard University Press, 2010.

Fisher, Edna M. "Prices of Sea Otter Pelts." *California Fish and Game* 27 (1941): 264.

Fisher, Robin. "The Northwest from the Beginning of Trade with Europeans to the 1880s." In *The Cambridge History of Native Peoples of the Americas, Vol. I, North America, Part 2*, edited by Bruce G. Trigger and Wilcomb E. Washburn, 117–82. New York: Cambridge University Press, 1996.

Fisher, Robin, and Hugh Johnston, eds. *From Maps to Metaphors: The Pacific World of George Vancouver*. Vancouver: University of British Columbia Press, 1993.

Forest, Marguerite S. E. "Searching for Sea Otters." *We Proceeded On* 33, no. 3 (August 2007): 18–27.

Freeman, Donald. *The Pacific*. London: Routledge, 2010.

Frost, Kathryn J. "Crosscurrents and Deep Water: Alaska's Marine Mammals." In *The Harriman Alaska Expedition Retraced: A Century of Change, 1899–2001*, edited by Thomas L. Litwin, 171–80. Piscataway NJ: Rutgers University Press, 2005.

Frost, Orcutt. *Bering: The Russian Discovery of America*. New Haven CT: Yale University Press, 2003.

Furstenberg, François. "The Significance of the Trans-Appalachian Frontier in Atlantic History." *American Historical Review* 113, no. 3 (June 2008): 647–77.

Galaup, Jean-François. *Voyages and Adventures of La Perouse*. Translated by Julius S. Gassner. Honolulu: University of Hawaii Press, 1969.

Gallo-Reynoso, Juan-Pablo. "Status of Sea Otters (*Enhydra lutris*) in Mexico." *Marine Mammal Science* 13, no. 2 (April 1997): 332–40.

Gibson, Arrell Morgan, and John S. Whitehead. *Yankees in Paradise: The Pacific Basin Frontier*. Albuquerque: University of New Mexico Press, 1993.

Gibson, James R. "The Exploration of the Pacific Coast." In *North American Exploration, Volume 2: A Continent Defined*, edited by John Logan Allen, 328–96. Lincoln: University of Nebraska Press, 1997.

———. "Nootka and Nutria: Spain and the Maritime Fur Trade of the Northwest Coast." In *Malaspina '92: I jornadas internacionales*, 137–60. Cádiz: Real Academia Hispano-Americana, 1994.

———. *Otter Skins, Boston Ships, and China Goods: The Maritime Fur Trade of the Northwest Coast, 1785–1841*. Seattle: University of Washington Press, 1992.

———. "Sitka–Kyakhta versus Sitka–Canton: Russian America and the China Market." *Pacifica* 2 (November 1990): 35–79.

Giesecke, E. W. "Unlikely Partners: Bostonians, Russians, and Kodiaks Sail the Pacific Coast Together, 1800–1810." *Mains'l Haul: A Journal of Pacific Maritime History* 43, nos. 3 and 4 (Summer/Fall 2007): 34–69.

Gilbert, Bil. "Dept. of Otter Confusion." *Sports Illustrated*, July 26, 1976, 63–66, 69–72.

Glover, James M. "Sweet Days of a Naturalist: Olaus Murie in Alaska, 1920–1926." *Forest and Conservation History* 36, no. 3 (1992): 132–40.

Goetzmann, William H., and Kay Sloan. *Looking Far North: The Harriman Expedition to Alaska, 1899*. Princeton NJ: Princeton University Press, 1982.

Gough, Barry. *Fortune's a River: The Collision of Empires in Northwest America*. Madeira Park BC: Harbour Publishing, 2007.

Graebner, Norman A. *Empire on the Pacific: A Study in American Continental Expansion*. Claremont CA: Regina Books, 1983.

Grinev, Andrei V. "Why Russia Sold Alaska: The View from Russia." Translated by Richard L. Bland. *Alaska History* 19 (2004): 1–22.

Hackel, Steven W. *Children of Coyote, Missionaries of Saint Francis: Indian-Spanish Relations in Colonial California, 1769–1850*. Chapel Hill: University of North Carolina Press, 2005.

Hardee, Jim. "Soft Gold: Animal Skins and the Early Economy of California." In *Studies in Pacific History: Economics, Politics, and Migration*, edited by Dennis O. Flynn, Arturo Giraldez, and James Sobredo, 22–39. Farnham: Ashgate Publishing Limited, 2002.

Hatch, David R. "Elakha: Sea Otters, Native People, and the European Colonization of the North Pacific." In *Changing Landscapes: "Sustaining Traditions," Proceedings of the 5th and 6th Annual Coquille Cultural Preservation Conferences*, edited by Donald B. Ivy and R. Scott Byram. North Bend OR: Coquille Indian Tribe, 2002.

Hattori, Kaoru, Ichiro Kawabe, Ayako W. Mizuno, and Noriyuki Ohtaishi. "History and Status of Sea Otters, *Enhydra lutris* along the Coast of Hokkaido, Japan." *Mammal Study* 30 (2005): 41–51.

Heffer, Jean. *The United States and the Pacific: History of a Frontier*. Translated by W. Donald Wilson. Notre Dame, IN: University of Notre Dame Press, 2002.

Hellyer, Robert. "The West, the East, and the Insular Middle: Trading Systems, Demand, and Labour in the Integration of the Pacific, 1750–1875." *Journal of Global History* 8, no. 3 (November 2013): 391–413.

Hinckley, Ted C. *The Americanization of Alaska, 1867–1897*. Palo Alto CA: Pacific Books, 1972.

Holmes, Cynthia. *Otters Holding Hands*. YouTube video, 01:40. Posted [March 2007]. https://www.youtube.com/watch?v=epUk3T2Kfno.

Hooper, C. L. *A Report on the Sea-Otter Banks of Alaska*. Washington DC: Government Printing Office, 1897.

Igler, David. "Diseased Goods: Global Exchanges in the Eastern Pacific Basin, 1770–1850." *American Historical Review* 109, no. 3 (June 2004): 693–719.

———. "Exploring the Concept of Empire in Pacific History: Individuals, Nations, and Ocean Space prior to 1850." *History Compass* 12, no. 11 (2014): 879–87.

———. *The Great Ocean: Pacific Worlds from Captain Cook to the Gold Rush*. New York: Oxford University Press, 2013.

———. "Hardly Pacific: Violence and Death in the Great Ocean." *Pacific Historical Review* 84, no. 1 (February 2015): 1–18.

———. "The Northeastern Pacific Basin: An Environmental Approach to Seascapes and Littoral Places." In *A Companion to American Environmental History*, edited by Douglas Cazaux Sackman. Hoboken NJ: Wiley-Blackwell, 2010.

Inglis, Robin. "The Effect of Laperouse on Spanish Thinking about the Northwest Coast." In *Spain and the North Pacific Coast: Essays in Recognition of the Bicentennial* of the Malaspina Expedition 1791–1792, edited by Robin Inglis, 46–52. Vancouver: Vancouver Maritime Museum, 1992.

Irish, Ann B. *Hokkaido: A History of Ethnic Transition and Development on Japan's Northern Island*. Jefferson NC: McFarland and Company, 2009.

Ivashchenko, Yulia, and Phil Clapham. "Soviet Illegal Whaling in the North Pacific: Reconstructing the True Catches." *NOAA Fisheries Quarterly Research Reports and Activities* (October–November–December 2012), https://www.afsc.noaa.gov/quarterly/ond2012/divrptsNMML2.htm.

Jackson, C. Ian. "The Stikine Territory Lease and Its Relevance to the Alaska Purchase." *Pacific Historical Review* 36, no. 3 (August 1967): 289–306.

Jewitt, John R. *A Narrative of the Adventures and Sufferings of John R. Jewitt; Only Survivor of the Ship Boston, during a Captivity of Nearly Three Years among the Savages of Nootka Sound: With an Account of the Manners, Mode of Living, and Religious Opinions of the Natives*. Middletown CT, 1815.

Jones, Robert. "Present Status of the Sea Otter in Alaska." In *Transactions of the Sixteenth North American Wildlife Conference*, edited by E. Quee, 376–83. Washington DC: Wildlife Management Institute, 1951.

Jones, Ryan Tucker. *Empire of Extinction: Russians and the North Pacific's Strange Beasts of the Sea, 1741–1867*. New York: Oxford University Press, 2014.

———. "Running into Whales: The History of the North Pacific from below the Waves." *American Historical Review* 118, no. 2 (April 2013): 349–77.

Jones, Terry L., Brendan J. Culleton, Shawn Larson, Sarah Mellinger, and Judith Porcasi. "Toward a Prehistory of the Southern Sea Otter (*Enhydra lutris nereis*)." In *Human Impacts on Seals, Sea Lions, and Sea Otters: Integrating Archaeology and Ecology in the Northeast Pacific*, edited by Todd J. Braje and Torben C. Rick, 243–71. Berkeley: University of California Press, 2011.

Jordan, David Starr. "Colonial Lessons of Alaska." *Atlantic Monthly*, November 1898, 581.

———, ed. *Seal and Salmon Fisheries and General Resources of Alaska*. Vol. 1. Washington DC: Government Printing Office, 1898.

Kahn, Ava F., ed. *Jewish Voices of the California Gold Rush: A Documentary History, 1849–1880*. Detroit: Wayne State University Press, 2002.

Kawata, Ken. "Zoological Gardens of Japan." In *Zoo and Aquarium History: Ancient Animal Collections to Zoological Gardens*, edited by Vernon N. Kisling Jr., 295–330. Boca Raton FL: CRC Press, 2001.

Keiner, Christine. "How Scientific Does Marine Environmental History Need to Be?" *Environmental History* 18 (January 2013): 111–20.

Kenyon, Karl W. *The Sea Otter in the Eastern Pacific Ocean*. New York: Dover Publications, 1975.

Kimberly, Jane Merritt. "50 Years and More in Santa Barbara." *Noticias: Quarterly Magazine of the Santa Barbara Historical Society* 34, no. 3 (Autumn 1988): 51–62.

"Kimberly, Martin Morse." *Islapedia*, http://islapedia.com/index.php?title=KIMBERLY,_Martin_Morse.

King, Robert J. "The Mulkovsky Expedition and Catherine II's North Pacific Empire." *Australian Slavonic and East European Studies* 21, nos. 1 and 2 (2007): 97–122.

Kinney, D. J. "The Otters of Amchitka: Alaskan Nuclear Testing and the Birth of the Environmental Movement." *Polar Journal* 2, no. 2 (2012): 291–311.

Kirkpatrick, Charles M., Donald E. Stullken, and Robert D. Jones Jr. "Notes on Captive Sea Otters." *Arctic* 8, no. 1 (1955): 46–59.

Koerper, Henry C. "Two Sea Otter Effigies and Three Pinniped Effigies: Illustrations, Descriptions, and Discussions." *Pacific Coast Archaeological Society Quarterly* 45, nos. 1 and 2 (August 2011): 101–22.

Kohlhoff, Dean W. *Amchitka and the Bomb: Nuclear Testing in Alaska.* Seattle: University of Washington Press, 2002.

Kornev, S. I., and S. M. Korneva. "Historical Trends in Sea Otter Populations of the Kuril Islands and South Kamchatka." In *Alaska Sea Otter Research Workshop: Addressing the Decline of the Southwestern Alaska Sea Otter Population*, edited by Daniela Maldini, Donald Calkins, Shannon Atkinson, and Rosa Meehan, 21–23. Fairbanks: Alaska Sea Grant College Program, University of Alaska Fairbanks, 2004.

Krasheninnikov, S. P. *The History of Kamtschatka and the Kurilski Islands, with the Countries Adjacent.* Translated by James Grieve. Chicago: Quadrangle Books, 1962.

Kushner, Howard J. *Conflict on the Northwest Coast: American-Russian Rivalry in the Pacific Northwest, 1790–1867.* Westport CT: Greenwood Press, 1975.

Lamb, W. Kaye, and Tomas Bartroli. "James Hanna and John Henry Cox: The First Maritime Fur Trader and His Sponsor." *BC Studies* 84 (Winter 1989–90): 3–36.

Landgon, Stephen L. "Efforts at Humane Engagement: Indian-Spanish Encounters in Bucareli Bay, 1779." In *Enlightenment and Exploration in the North Pacific 1741–1805*, edited by Stephen Haycox, James Barnett, and Caedmon Liburd, 187–97. Seattle: University of Washington Press, 1997.

Langsdorff, Georg Heinrich Von. *Remarks and Observations on a Voyage around the World from 1803–1807.* Vol. 2. Translated and annotated by Victoria Joan Moessner, edited by Richard A. Pierce. Kingston, ON: Limestone Press, 1993.

Larson, Henrietta M. "William Sturgis, Merchant and Investor." *Bulletin of the Business Historical Society* 9, no. 5 (October 1935): 76–77.

Larson, Shawn E., and James L. Bodkin. "The Conservation of Sea Otters: A Prelude." In *Sea Otter Conservation*, edited by Shawn E. Larson, James L. Bodkin, and Glenn R. VanBlaricom, 1–17. London: Elsevier, 2015.

Lavigne, David M., Victor B. Scheffer, and Stephen R. Kellert. "The Evolution of North American Attitudes toward Marine Mammals." In *Conservation and Management of Marine Mammals*, edited by John R. Twiss and Randall R. Reeves. Washington DC: Smithsonian Institution Press, 1999.

Lavrov, K. P. "Fur Treasures of the Commodore Islands." *Trans-Pacific* 5 (July–December 1921): 74.

Leatherwood, Stephen, Linda J. Harrington-Coulombe, and Carl L. Hubbs. "Relict Survival of the Sea Otter in Central California and Evidence of Its Recent Redispersal South of Point Conception." *Bulletin of Southern California Academy of Sciences* 77 (1978): 109–15.

Lech, Veronica, Matthew W. Betts, and Herbert D. G. Maschner. "An Analysis of Seal, Sea Lion, and Sea Otter Consumption Patterns on Sanak Island, Alaska." In *Human Impacts on Seals, Sea Lions, and Sea Otters: Integrating Archaeology and Ecology in the Northeast Pacific*, edited by Todd J. Braje and Torben C. Rick. Berkeley: University of California Press, 2011.

Lee, Molly. "Context and Contact: The History and Activities of the Alaska Commercial Company, 1867–1900." In *Catalogue Raisonnée of the Alaska Commercial Company Collection, Phoebe Apperson Hearst Museum of Anthropology*, edited by Nelson H. H. Graburn, Molly Lee, and Jean-Loup Rousselot, 19–38. Berkeley: University of California Press, 1996.

Lensen, George Alexander. *The Russian Push toward Japan: Russo-Japanese Relations, 1697–1875*. Princeton NJ: Princeton University Press, 1959.

Lensink, C. J. "The History and Status of Sea Otters in Alaska." Ph.D. dissertation, Purdue University, 1962.

Leon, Vicki. *A Raft of Sea Otters: The Playful Life of a Furry Survivor*. Montrose CA: London Town Press, 2005.

Lisiansky, Urey. *A Voyage round the World, in the Years 1803, 5, & 6* . . . London: John Booth et al., 1814.

Loring, Charles G. "Memoir of William Sturgis." *Proceedings of the Massachusetts Historical Society* 7 (1864): 458.

Macdonald, Augustin S. *Pacific Pelts: Sea Otters Choose California Coast*. Oakland CA: Unknown publisher, 1938.

Mackie, Richard Somerset. *Trading beyond the Mountains: The British Fur Trade on the Pacific, 1793–1843*. Vancouver: University of British Columbia Press, 1997.

Malloy, Mary. *Boston Men on the Northwest Coast: The American Fur Trade, 1788–1884*. Fairbanks: University of Alaska Press, 1998.

———. *"A Most Remarkable Enterprise": Lectures on the Northwest Coast Trade and Northwest Coast Indian Life by Captain William Sturgis*. Marstons Mills MA: Parnassus Imprints, 2000.

Mapp, Paul W. *The Elusive West and the Contest for Empire, 1713–1763*. Chapel Hill: University of North Carolina Press, 2013.

Marakov, S. V. "A Natural Monument or a Commercial Species? The Future of the Commander Islands Sea Otters." *Priroda* 52, no. 11 (1963), http://www.dfo-mpo.gc.ca/Library/30460.pdf.

Matsuda, Matt K. *Pacific Worlds: A History of Seas, Peoples, and Cultures*. New York: Cambridge University Press, 2012.

Matthews, Owen. *Glorious Misadventures: Nikolai Rezanov and the Dream of a Russian America*. New York: Bloomsbury, 2013.

McCracken, Harold. *Hunters of the Stormy Sea*. New York: Doubleday and Company, 1957.

McDougall, Walter A. *Let the Sea Make a Noise . . . : A History of the North Pacific from Magellan to MacArthur*. New York: Basic Books, 1993.

McKechme, Iain, and Rebecca J. Wigen. "Toward a Historical Ecology of Pinniped and Sea Otter Hunting Traditions on the Coast of Southern British Columbia." In *Human Impacts on Seals, Sea Lions, and Sea Otters: Integrating Archaeology and Ecology in the Northeast Pacific*, edited by Todd J. Braje and Torben C. Rick, 129–66. Berkeley: University of California Press, 2011.

Miller, Gwenn A. "Russian Routes: Kamchatka to Kodiak Island." *Common-place* 5, no. 2 (January 2005), http://www.common-place.org/vol-05/no-02/miller/index.shtml.

Miller, Robert J. "The International Law of Discovery: Acts of Possession on the Northwest Coast of North America." In *Arctic Ambition: Captain Cook and the Northwest Passage*, edited by James K. Barnett and David L. Nicandri, 191–209. Anchorage: Anchorage Museum, 2015.

Montanari, Shaena. "Rare Otter Fossil Found in the Mexican Desert." *National Geographic*, June 14, 2017, http://news.nationalgeographic.com/2017/06/otters-fossils-americas-mexico-paleontology-science.

Moss, Madonna L., and Robert J. Losey. "Native American Use of Seals, Sea Lions, and Sea Otters in Estuaries of Northern Oregon and Southern Washington." In *Human Impacts on Seals, Sea Lions, and Sea Otters: Integrating Archaeology and Ecology in the Northeast Pacific*, edited by Todd J. Braje and Torben C. Rick, 167–95. Berkeley: University of California Press, 2011.

Muller, Adolph. "Appendix V: The Fur Trade at San Francisco, California." In *Report of the Commissioner of Indian Affairs, Made to the Secretary of the Interior, for the Year 1869*. Washington DC, 1869.

Murphy, Robert. "Tales of the Sea Otter." *Saturday Evening Post*, December 10, 1949, 30, 128–30.

Murray, Michael J. "Veterinary Medicine and Sea Otter Conservation." In *Sea Otter Conservation*, edited by Shawn E. Larson, James L. Bodkin, and Glenn R. VanBlaricom, 159–95. London: Elsevier, 2015.

Murray, Peter. *Sea Otters*. Chanhassen MN: Child's World, 2001.

———. *The Vagabond Fleet: A Chronicle of the North Pacific Sealing Schooner Trade*. Victoria BC: Sono Nis Press, 1988.

Mutual of Omaha's Wild Kingdom: World of the Sea Otter. YouTube video, 22:15. Posted [June 2009]. https://www.youtube.com/watch?v=lECZuIAiJVI.

Nance, Susan, ed. *The Historical Animal*. Syracuse NY: Syracuse University Press, 2015.

Newland, Renee. "Bixby Creek Bridge." Monterey County Historical Society, http://www.mchsmuseum.com/bixbycr.html.

Nichol, Linda M. "Conservation in Practice." In *Sea Otter Conservation*, edited by Shawn E. Larson, James L. Bodkin, and Glenn R. VanBlaricom, 369–93. London: Elsevier, 2015.

Nickerson, Roy. *Sea Otters: A Natural History and Guide*. San Francisco: Chronicle Books, 1989.

NNSANevada. *The Warm Coat—1968–1969*. YouTube video, 14:01. Posted [September 2015]. https://www.youtube.com/watch?v=X6ldPRpHrrQ.

Nunis, Doyce B., Jr. "Alta California's Trojan Horse: Foreign Immigration." In *Contested Eden: California before the Gold Rush*, edited by Ramon A. Gutierrez and Richard J. Orsi, 299–330. Berkeley: University of California Press, 1998.

Ogden, Adele. *The California Sea Otter Trade, 1784–1848*. Berkeley: University of California Press, 1941.

———. "The Californias in Spain's Pacific Otter Trade, 1775–1795." *Pacific Historical Review* 1, no. 1 (December 1932): 444–69.

Owens, Kenneth N., with Alexander Yu. Petrov. *Empire Maker: Aleksandr Baranov and Russian Colonial Expansion into Alaska and Northern California*. Seattle: University of Washington Press, 2015.

Oyer, P. H. "Sea Otters near Monterey." *California Fish and Game* 3, no. 2 (1917): 88.

Palumbi, Stephen R., and Carolyn Sotka. *The Death and Life of Monterey Bay: A Story of Revival*. Washington DC: Island Press, 2011.

Peebles, Jonathan. *The Undersea World of Jacques Cousteau: The Unsinkable Sea Otter!* YouTube video, 48:37. Posted [October 2016]. https://www.youtube.com/watch?v=Zh1larkJeXw.

Phelps, William Dane. "Solid Men of Boston in the Northwest." In *Fur Traders from New England: The Boston Men in the North Pacific, 1787–1800*, edited by Briton C. Busch and Barry M. Gough, 31–83. Spokane WA: Arthur H. Clark Company, 1997.

Phillipi, Donald L. *Songs of Gods, Songs of Humans: The Epic Tradition of the Ainu*. Princeton NJ: Princeton University Press, 1979.

Pletcher, David M. *The Diplomacy of Involvement: American Economic Expansion across the Pacific, 1784–1900.* Columbia: University of Missouri Press, 2001.

Plummer, Katherine. *The Shogun's Reluctant Ambassadors: Japanese Sea Drifters in the North Pacific.* Portland: Oregon Historical Society, 1991.

Pomeranz, Kenneth, and Steven Topik. *The World That Trade Created: Society, Culture, and the World Economy, 1400–Present* 2nd ed. Armonk NY: M. E. Sharpe, 2006.

Rauzon, Mark. "Memories." *Marine Mammal Science* 23, no. 4 (2007): 998.

Ravalli, Richard. "Graphing the Sea Otter Hunt." Alaska Historical Society Blog, November 6, 2015, http://alaskahistoricalsociety.org/graphing-the-sea-otter-hunt.

———. "Grounds of Our Claim: William Sturgis and Commercial Diplomacy on the Northwest Coast." *Columbia: The Magazine of Northwest History* 28, no. 1 (Spring 2014): 14–20.

———. "Sea Otter Aesthetics and Popular Culture." In *Animals in Human Society: Amazing Creatures Who Share Our Planet,* edited by Daniel Moorehead, 93–103. Lanham MD: University Press of America, 2016.

Ray, Arthur J. "Some Conservation Schemes of the Hudson's Bay Company, 1821–1850: An Examination of the Problems of Resource Management in the Fur Trade." In *The American Environment: Interpretations of Past Geographies,* edited by Lary M. Dilsaver and Craig E. Colten, 33–50. Lanham MD: Rowman and Littlefield, 1992.

Ray, G. Carlton, and Frank M. Potter Jr. "The Making of the Marine Mammal Protection Act of 1972." *Aquatic Mammals* 37, no. 4 (2011): 522–52.

Reid, Joshua L. *The Sea Is My County: The Maritime World of the Makahs.* New Haven CT: Yale University Press, 2015.

Rick, Torben C., Jon M. Erlandson, Todd J. Braje, James A. Estes, Michael H. Graham, and Rene L. Vellanoweth. "Historical Ecology and Human Impacts on Coastal Ecosystems of the Santa Barbara Channel Region, California." In *Human Impacts on Ancient Marine Ecosystems: A Global Perspective,* edited by Torben C. Rick and Jon M. Erlandson, 77–101. Berkeley: University of California Press, 2008.

Rogers, Eugene F. "Memoirs." *Noticias: Quarterly Bulletin of the Santa Barbara Historical Society* 26 (1980): 47–50.

Ronda, James P. *Astoria and Empire.* Lincoln: University of Nebraska Press, 1990.

———. "The Education of an Empire Builder: John Jacob Astor and the World of the Columbia." *Columbia: The Magazine of Northwest History* 11, no. 3 (Fall 1997): 1–7.

Salomon, Anne K., Kii'iljuus Barb J. Wilson, Xanius Elroy White, Nick Tanape Sr., and Tom Mexsis Happynook. "First Nations Perspectives on Sea Otter Conservation in British Columbia and Alaska: Insights into Coupled Human-Ocean Systems." In *Sea Otter Conservation,* edited by Shawn E. Larson, James L. Bodkin, and Glenn R. VanBlaricom, 301–31. London: Elsevier, 2015.

Sandos, James A. *Converting California: Indians and Franciscans in the Missions.* New Haven CT: Yale University Press, 2004.

Sato, Kumito. "From Hello Kitty to Cod Roe Kewpie: A Postwar Cultural History of Cuteness in Japan." *Education about Asia* 14, no. 2 (Fall 2009): 38–42.

Saunt, Claudio. *West of the Revolution: An Uncommon History of 1776.* New York: W. W. Norton and Company, 2014.

Scammon, Charles M. *The Marine Mammals of the North-western Coast of North America.* Riverside CA: Manessier, 1969.

Scheffer, Victor. "The Last of the Sea Otter Hunters." *Columbia: The Magazine of Northwest History* 13, no. 4 (Winter 1999–2000): 15.

———. "The Sea Otter on the Washington Coast." *Pacific Northwest Quarterly* 3 (1940): 370–88.

Schlesinger, Jonathan. *A World Trimmed with Fur: Wild Things, Pristine Places, and the Natural Fringes of Qing Rule.* Palo Alto CA: Stanford University Press, 2017.

Schwartz, Harvey. "Fort Ross, California: Imperial Russian Outpost on America's Western Frontier, 1812–1841." *Journal of the West* 18, no. 2 (1979): 35–48.

Scofield, John. *Hail, Columbia: Robert Gray, John Kendrick and the Pacific Fur Trade.* Portland: Oregon Historical Society Press, 1993.

"Sea Otters." In *Islapedia,* http://www.islapedia.com/index.php?title=SEA_OTTERS.

"Sea Otters in Alaska." In *Harriman Expedition Retraced: A Century of Change,* http://www.pbs.org/harriman/1899/seaotters.html.

Sharpe, Howard Granville. "The Discovery of the 'Extinct' Sea Otter." *Friends of the Sea Otter,* http://www.seaotters.org/Otters/index.cfm?DocID=8.

Shaw, George. *Musei Leveriani Explication, Anglica et Latina.* London: James Parkinson, 1792.

Shelikhov, Grigorii I. *A Voyage to America: 1783–1786.* Translated by Marina Ramsay, edited by Richard A. Pierce. Kingston ON: Limestone Press, 1981.

Shubin, Valery O. "Russian Settlements in the Kurile Islands in the 18th and 19th Centuries." In *Russia in North America: Proceedings of the 2nd International Conference on Russian America,* edited by Richard A. Pierce. Kingston ON: Limestone Press, 1990.

Silverstein, Alvin, Virginia Silverstein, and Robert Silverstein. *The Sea Otter.* Brookfield CT: Millbrook Press, 1995.

Simpson, George. *An Overland Journey round the World, during the Years 1841 and 1842.* Philadelphia: Lea and Blanchard, 1847.

Sloan, N. A., and Lyle Dick. *Sea Otters of Haida Gwaii: Icons in Human-Ocean Relations.* Skidegate: Haida Gwaii Museum, 2012.

Snow, H. J. *In Forbidden Seas: Recollections of Sea-Otter Hunting in the Kurils.* London: Edward Arnold, 1910.

Solovjova, Katerina, and Aleksandra Vovnyanko. "The Rise and Decline of the Lebedev-Lastochkin Company: Russian Colonization of South Central Alaska, 1787–1798." *Pacific Northwest Quarterly* 90, no. 4 (Fall 1999): 191–205.
Sowards, Adam M., ed. *United States West Coast: An Environmental History*. Santa Barbara CA: ABC-CLIO, 2007.
Steller, Georg Wilhelm. *De bestiis marinis, or, The Beasts of the Sea*. Edited by Paul Royster and translated by Walter Miller and Jennie Emerson Miller. Zea E-Books, 2011. https://digitalcommons.unl.edu/zeabook/1/.
———. *History of Kamchatka*. Edited by Marvin W. Falk, translated by Margritt Engel and Karen Willmore. Fairbanks: University of Alaska Press, 2003.
———. *Journal of a Voyage with Bering, 1741–1742*. Edited by O. W. Frost, translated by Margritt A. Engel and O. W. Frost. Stanford CA: Stanford University Press, 1988.
Stephan, John. *The Kuril Islands: Russo-Japanese Frontier in the Pacific*. London: Oxford University Press, 1974.
Stevens, Sadie S., John F. Organ, and Thomas L. Serfass. "Otters as Flagships: Social and Cultural Considerations." *Proceedings of Xth International Otter Colloquium*, IUCN *Otter Specialist Group Bulletin* 28, A (2011), http://iucnosg.org/Bulletin/Volume28A/Stevens_et_al_2011.html#Fig1.
Stolberg, Eva-Maria. "Interracial Outposts in Siberia: Nerchinsk, Kiakhta, and the Russo-Chinese Trade in the Seventeenth/Eighteenth Centuries." *Journal of Early Modern History* 4, nos. 3 and 4 (2000): 322–36.
Stone, Ian R. "Hunting Marine Mammals for Profit and Sport: H. J. Snow in the Kuril Islands and the North Pacific, 1873–1896." *Polar Record* 41, no. 216 (2005): 47–55.
Sturgis, William. "Examination of the Russian Claims to the Northwest Coast of America." *North American Review* 15 (October 1822): 370–401.
Takahashi, Chikashi. "Inter-Asian Competition in the Fur Market in the Eighteenth and Nineteenth Centuries." In *Intra-Asian Trade and the World Market*, edited by A. J. H. Latham and Heita Kawakatsu, 37–45. London: Routledge, 2006.
Taylor, George P. "Spanish-Russian Rivalry in the Pacific, 1769–1820." *Americas* 15, no. 2 (October 1958): 109–27.
Taylor, Lawrence. "The Mining Boom in Baja California from 1850 to 1890 and the Emergence of Tijuana as a Border Community." *Journal of the Southwest* 43 (2001): 463–92.
Tezuka, Kaoru. "Ainu Sea Otter Hunting from the Perspective of Sino-Japanese Trade." In *Human-Nature Relations and the Historical Backgrounds of Hunter-Gatherer Cultures in Northeast Asian Forests*, edited by Shiro Sasaki, 117–31. Osaka: National Museum of Ethnology, 2009.
———. "Long Distance Trade Networks and Shipping in the Ezo Region." *Arctic Anthropology* 35, no. 1 (1998): 350–60.

Tikhmenev, P. A. *A History of the Russian-American Company*. Translated and edited by Richard A. Pierce and Alton Donnelly. Seattle: University of Washington Press, 1978.

Torrubia, José. *The Muscovites in California*. Fairfield WA: Ye Galleon Press, 1996.

Tovell, Freeman M., trans. *Voyage to the Northwest Coast of America, 1792: Juan Francisco de la Bodega y Quadra and the Nootka Sound Controversy*. Norman OK: Arthur H. Clark Company, 2012.

Urness, Carol. "Russian Mapping of the North Pacific to 1792." In *Enlightenment and Exploration in the North Pacific 1741–1805*, edited by Stephen Haycox, James Barnett, and Caedmon Liburd, 132–46. Seattle: University of Washington Press, 1997.

VanBlaricom, Glenn. *Sea Otters*. St. Paul MN: Voyageur Press, 2001.

VanBlaricom, Glenn R., Traci F. Belting, and Lisa H. Triggs. "Sea Otters in Captivity: Applications and Implications of Husbandry Development, Public Display, Scientific Research and Management, and Rescue and Rehabilitation for Sea Otter Conservation." In *Sea Otter Conservation*, edited by Shawn E. Larson, James L. Bodkin, and Glenn R. VanBlaricom, 197–234. London: Elsevier, 2015.

———. "Synopsis of the History of Sea Otter Conservation in the United States." In *Sea Otter Conservation*, edited by Shawn E. Larson, James L. Bodkin, and Glenn R. VanBlaricom, 395–434. London: Elsevier, 2015.

VanBlaricom, G. R., and J. A. Estes, eds. *The Community Ecology of Sea Otters*. Berlin: Springer-Verlag, 1988.

Venegas, Miguel. *A Natural and Civil History of California*. Vol. 2. Ann Arbor MI: University Microfilms, 1966.

Vincenzi, Frank. "The Sea Otter (*Enhydra lutris*) at Woodland Park Zoo." In *The International Zoo Yearbook*, edited by Caroline Jarvis and Desmond Morris, 3:3, 27–29. London: Zoological Society of London, 1961.

Vinkovetsky, Ilya. *Russian America: An Overseas Colony of a Continental Empire, 1804–1867*. New York: Oxford University Press, 2011.

Walker, Brett. *Conquest of Ainu Lands: Ecology and Culture in Japanese Expansion, 1590–1800*. Berkeley: University of California Press, 2001.

Weiner, Douglas R. *A Little Corner of Freedom: Russian Nature Protection from Stalin to Gorbachev*. Berkeley: University of California Press, 1999.

Wells, David N., ed. and trans. *Russian Views of Japan, 1792–1913: An Anthology of Travel Writing*. London: Routledge Curzon, 2004.

Weyler, Rex. *Greenpeace: How a Group of Ecologists, Journalists, and Visionaries Changed the World*. Emmaus PA: Rodale, 2004.

Williams, Glyn. "James Cook and the Northwest Passage: Approaching the Third Voyage." In *Arctic Ambitions: Captain Cook and the Northwest Passage*, edited by James K. Barnett and David L. Nicandri, 21–42. Anchorage: Anchorage Museum, 2015.

Williams, Perry, and Melvin Hooten. "The Extraordinary Return of Sea Otters to Glacier Bay." *The Conversation*, April 19, 2017, http://theconversation.com/the-extraordinary-return-of-sea-otters-to-glacier-bay-74909.

Woolfenden, John, and Amelie Elkinton. *Cooper: Juan Bautista Rogers Cooper, Sea Captain, Adventurer, Ranchero, and Early California Pioneer, 1791–1872*. Pacific Grove CA: Boxwood Press, 1983.

Wright, Miranda. "The Sea Otter Industry in the Eastern Aleutians, 1867–1911." In *The History and Ethnohistory of the Aleutians East Borough*, edited by Lydia T. Black, 253–63. Kingston ON: Limestone Press, 1999.

Yamaura, Kiyoshi. "The Sea Mammal Hunting Cultures of the Okhotsk Sea with Special Reference to Hokkaido Prehistory." *Arctic Anthropology* 35, no. 1 (1998): 321–34.

Zilberstein, Anya. "Objects of Distant Exchange: The Northwest Coast, Early America, and the Global Imagination." *William and Mary Quarterly* 64, no. 3 (July 2007): 591–620.

INDEX

Page numbers in italic indicate illustrations.

abalone fishing, 111–12
Adams, John Quincy, 69, 70–71, 151n83
Ainu (Kuril Islanders): diet of, 5; heritage of, 4–5; hunting methods of, 5, 11–12; impacts of fur trade on, 3, 8, 13; under Japanese colonization, 14–15, 16, 144n39; otters in mythologies of, xxii; and Russian fur traders, 1–2, 16; trade networks of, 4–8; tribute (*yasak*), 2, 9, 12, 40; and violence during fur trade, 1–2, 144n39. *See also* Kuril Islands
Alaska: factors in otter survival in, 90; Harriman expedition to, 92–93; hunting regulations in, 113–14; otter hunting in, 27–29, 41–44, 52, 84–88, 113–14, 154n44; otter population levels in, 39–40, 98, 113, 118–19, 125–26; Russian exploration of, 25, 27–29; translocation efforts in, 125–26; U.S. purchase of, 83–84; U.S.-Russian geopolitical conflicts over, 67–71. *See also* Aleutian Islands; Kodiak Island
Alaska Commercial Company, 83–87, 93, 94
Alaska Purchase, 83–84
Albatross (ship), 65
Aleutian Islands: atomic testing at, 106, 107–9; conservation efforts at, 45–46, 105–9; factors in otter survival in, 89–90; otter hunting in, 26–28, 86, 89–90; otter population levels in, 39–43, 45–46, 86, 98, 118–19; prehistoric hunting in, xxi; Russian conservation efforts in, 45–46; violent encounters in, 2; during World War II, 105–6. *See also* Alaska
Aleutian Islands Refuge, 97
Aleuts (indigenous people): canoes used by, *18*, 146n20; hostage taking, 27; hunting in California, 63–64; hunting in the Kurils, 18; hunting methods of, 28; impacts of fur trade on, 27, 84, 87, 96; marriage to non-Native hunters, 87; otters in worldview of, xxii–xxiii; role in otter hunting, 28, 38–39, 44–45, 63–64, 87–88; role in otter population declines, 40, 44; violence during fur trade, 27–28
Alexander Archipelago, 44, 45
Alexander I (tsar), 15, 16, 69
Alta California: American fur traders in, 63–64; otter population levels in, 23; Spanish colonization of, 30–33; Spanish fur trading in, 35
Alutiiq (Kodiak Islanders): role in maritime fur trade, 28, 44–45; role in otter decline, 44; Russian treatment of, 34

179

Amchitka Island (Aleutian Islands): end of hunting at, 88; impact of World War II on, 105–6; Japanese shipwrecked on, 7; U.S. atomic tests at, 104, 106, 107–9
American fur traders: in California, 63–65, 73–76; in the China trade, 53–54; in the Hawaiian Islands, 61–63; in the Kuril Islands, 88–90; in the maritime fur trade, 47–50, 57–59; in the opium trade, 79–81; outposts of, 65–67; role in geopolitical conflicts, 71–73; role in otter population levels, 73–76; and the Russian contract system, 63–65; and the Russian *ukase*, 67–71; violence committed by, 59–61. *See also* United States
Ames, Jack, 100
Andreanov Islands, 40, 45
Ángeles, Jerónimo de, 6
anthropomorphism: impact on conservation, xxiii, 109; in indigenous mythologies, xxii–xxiii; in otter descriptions, xix, 12, 107, 109, 111, 119–20
aquaria, 104, 114–17
Arguello, Luis Antonio, 73
Astor, John Jacob, xv, 65–66, 67–68, 151n76
Astoria settlement, 49, 65–67, 68
Atlasov, Vladimir, 10–11
Atomic Energy Commission (U.S.), 108, 109
atomic tests, 104, 106, 107–9
Audubon, John Woodhouse, 75, 76

baidarkas (skin-constructed kayaks), 28, 146n20
Baja California: extirpation of sea otters in, 83, 92; otter hunting in, 35, 63–64, 77, 91; otter population levels in, 75; under Spain, 21, 35
Ballachey, Brenda E., 113
Bancroft, John, 72, 74
Baranov, Aleksandr, 14, 38, 41, 63, 67, 69
Barthélemy de Lesseps, Jean-Baptiste, 7–8

Basov, Emel'ian, 13, 26
Batten, B. T., 118
The Beasts of the Sea (Steller), xx, 12–13
beavers, 50, 67, 68, 71, 72, 126
Benton, Thomas Hart, 68, 151n76
Bering, Vitus, 11, 19, 24–26, 29, 30
Bering Island, 25–26
Berkh, Vasilii Nikolaevich, 28
De bestiis marinis (Steller), xx, 12–13
Betts, Matthew W., xxi
Billings, Joseph, 37–38, 41
Bixby Creek (CA), 99, 101, 158n52
Black, Lydia T., 27, 34, 144n36, 145n11
Block, Michael D., 150n64
Bodega Bay (CA), 33, 63
Bodega y Quadra, Juan Francisco de la, 32–33, 57
Bodkin, James L., 113, 148n76
Boston (ship), 60
Boston Men. *See* American fur traders
British Columbia: extirpation of sea otters in, 83; prehistoric otter hunting in, xxii; violence during fur trade in, 44; wildlife regulations in, 114
British fur traders: in the China trade, 51, 52–53; in Fort Astoria, 66–67; in the Hawaiian Islands, 61–63; in the Kuril Islands, 88–90; in the maritime fur trade, 50–53, 61–63, 71–73; in the Nootka Controversy, 36, 38, 54–57; ships in California otter trade, 131–32, 135–39; trade agreements with Russia, 71–72; trade restrictions, 52–53
Broughton, William, 15
Bucareli Bay (AK), 33, 41
Bucich, Richard, 119
Busch, Briton C., 65
Bustamante y Guerra, José, 43–44

Calderón y Henríquez, Pedro, *31*
California: American fur traders in, 43, 50, 63–64, 65, 73–76; compared to

Alaska, 41–44; conservation efforts in, 23, 74, 111–12; decline of otter hunting in, 81–82; extirpation of sea otters in, 83, 92; gold rush in, 91; hunting regulations in, 97–98; otter population levels in, 23, 43–44, 73–76, 90–92, 99–101, 113; pelt exports from, 128–39, 147n53; rediscovery of otters in, 78–79; Ross settlement in, 37, 38–39, 45, 64, 68, 147n58; Russian contract system in, 63–64; Russian fur traders in, 38–39, 43, 50, 73; Russian-Spanish geopolitical tensions over, 21–22; ships in fur trade, 131–39; Spanish colonization of, 30–33; Spanish fur trading in, 35–37, 43–44; targeting of otter pups in, 152n103

The California Sea Otter Trade, 1784–1848 (Ogden), 127–39

Canada: border with U.S., 47–48, 67; fur trade in, 56, 71, 94–95; geopolitical conflict in, 94–95, 96; hunting prohibited by, 96, 98; wildlife regulations in, 114

Carlos III (king of Spain), 30–31

Carvalho, Diego, 6

Catalina Island (CA), 99

Catherine the Great (empress of Russia), 8, 37

Channel Islands: extermination of otters in, 92; otter hunting in, 77, 90–91; otter population levels in, 75; prehistoric otter hunting in, xxi

Chiang, Connie, 116

China trade: American participation in, 47, 53–54, 58, 59, 79–81; British participation in, 51, 52–53; demand for fur in, 4, 24, 28, 51; expansion of, 24, 79–81; increasing competition in, 143n27; Japanese exports in, 145n53; opium in, 79–81; Spanish participation in, 35, 36; trade networks with Ainu, 6; trade networks with Russia, 9–10, 153n8. *See also* fur trade

Chirikov, Aleksei, 24–25

Chumash (Channel Islanders), xxi

Clapham, Phil, 110

Clark, William, 151n76

Clayton, Daniel, 56, 60–61

Cochrane, John Dundas, 143n27

Colnett, James, 54–56

Columbia (ship), 54, 61

Columbia Department (Hudson's Bay Company), 71–73

Columbia River, 54, 65–69, 149n20

Commander Islands: captivity experiments in, 98; Native hunting in, 90; Russian conservation at, 90, 110; Russian hunting in, 26, 27

Congress (U.S.), 83, 112–13

conservation: in the Aleutian Islands, 45–46, 105–9; anthropomorphism's impact on, xxiii, 109; and atomic testing, 105–9; building public support, 111–12, 115; in California, 23, 74, 111–12; in Canada, 114; in the Commander Islands, 90, 110; factors in, 104–5, 111–12; focusing on otter aesthetic qualities, 112; for fur seals, 18, 45, 85–86, 93–97, 101; impact of otter cuteness on, 121; impacts of zoos and aquaria on, 117; in the Kuril Islands, 109–10, 145n55; of the Progressive Era, 93–98; Russian policies of, 18–19, 23, 40, 45–46, 84, 110, 145n55; sea otters as a flagship species, 121; successes of, 101, 104, 123; U.S. policies of, 85–86, 112–14

Cook, James, 2, 35, 38, 50–51

Cook, Warren L., 56, 147n53

Coolidge, Calvin, 125

Cooper, John Rogers, 73

Corney, Peter, 62–63

Cousteau, Jacques, 111

INDEX 181

Cox, John Henry, 51, 53
Coxe, William, 27, 28
cuteness (sea otters): in animal videos, 103–4, 122–23; benefits of, xxiii, 111, 121, 123; and consumerism, 116; contrasted with aggressiveness, 121–22; in general audience books, 119–20; problems with, 116–17, 120–21
Cygnet (schooner), 77–78, *80*, 86, 88

Daily, Marla, 91
Dana, Richard Henry, 65
Dashkov, Andrey, 67–68
Davis, Isaac, 62
Davydov, Gavril, 15
Del Bucchia, Dina, 123
Delisle, Joseph-Nicolas, 30
Dembei (Japanese castaway), 10–11
Dick, Lyle, xxiii
Discovery (ship), 50–51
diseases, 62, 101
Dixon, George, 52
Don't Make a Wave Committee, 107, 157n14
Door, Sullivan, 53
Dorsey, Kurkpatrick, 96
Douglas, William, 54
Dye, Job, 74, 152n100

East India Company, 51, 52, 55, 56, 79
Elliott, Henry Wood, 86, 90–91, 95, 97, 154n27
Emmons, George Thornton, 88
Empress of China (ship), 53
Endangered Species Act (1973), 112–13
English fur traders. *See* British fur traders
Engstrand, Iris H. W., 33
Enhydra lutris (sea otter). *See* sea otters
Enhydritherium, xvii
environmental activism, 107–8, 111–12, 118. *See also* conservation
Estes, James, xviii, 100, 109, 119, 141n9
Exxon Valdez oil spill, 103, 117–18

falconry, 7
Farris, Glenn, 127
firearms, 62, 64, 71, 86, 90, 95, 154n27
First Kamchatka Expedition, 24
fishing industries, 111–12, 114
Floyd, John, 68, 69, 151n76
Fox Islands, 40
France, 57, 135, 137
Friendly Cove (Vancouver Island), 54, 55
Friends of the Sea Otter (NGO), 111–12
Frost, Kathryn J., 93
fur (sea otters): demand for, 4, 6–7, 10, 17, 28; description of, xviii; indigenous use of, xxii–xxiii; price of, 7, 28, 58, 82
fur seals. *See* seals (fur seals)
Furstenberg, François, 49
fur trade: Alaska-California comparisons, 41–44; decline of, 82–83, 88, 90, 91–93, 153n16; early networks of, 4–8; eastern and western Pacific parallels, 22; economic importance of, 13, 56, 67; export figures in, 42, 128–39; geopolitical importance of, xiii, xiv–xv, 2–3, 13, 16, 36, 48–50, 56, 62, 64–65, 68–71, 94; hostage taking during, 27, 34, 59; hunting methods in, 5, 11–12, 28, 73–74, 82, 86, 89, 154n27, 154n44; illegal hunting in, 77–79, 89; impacts on indigenous people, 3, 8, 13, 27, 60–61, 62, 84, 87, 96; independent hunters in, 86–87, 91; Japanese networks in, 9; list of ships in, 131–39; Native hunters in, 27, 35, 44–45, 72, 82, 84, 86–88; pelagic hunting, 85–86, 94–96, 154n31; previous scholarship on, xiii–xiv; regulations regarding, 52–53, 87, 97–98, 113–14; role of competition in, 33; Russian-Chinese networks, 9–10; shifting markets for, 24, 79–81; Spanish participation in, 23, 30, 32,

34–37; targeting of otter pups in, 74, 152n103; types of animals hunted, 28; violence during, 1–2, 27–28, 33–34, 44, 51–52, 59–61. *See also* American fur traders; British fur traders; China trade; Russian fur traders

Galaup, Jean-François de, 7
Gale, William, 65, 150n64
Galvez, José de, 31
Gannett, Henry, 93
Gibson, Arrell Morgan, 79
Gibson, James R., 10, 54, 58, 71
Gilbert, Grove Karl, 93
ginseng (North American), 53
Glacier Bay (AK), 125–26
Glaciers and Glaciation (Gilbert), 93
gold rush, 91
Goolovnin, Vasilii Mikhailovich, 15–16
Gough, Barry M., 65
Gray, Robert, 54, 59–60, 61
great white sharks, xix, 101
Greenpeace Foundation, 107
Grover, Steve, 82

Haida Gwaii (Queen Charlotte Islands), xx–xxi, 32, 41, 52, 54
Hanna, James, 51–52, 61
Harriman, E. H., 92–93, 125
Hart, Thomas, 68
Hatfield, Brian, 100
Hawaiian Islands, 50, 51, 59, 61–63, 136–37
hawks, 7
Hellyer, Robert, 3
Hezeta, Bruno de, 32–33, 149n20
Hiroshige, Utagawa, III, 18
Hokkaido: in the China trade, 4, 6–7; hawk industry on, 7; impacts of trade on, 8; Japanese control of, 14–15; in Japanese trade networks, 6–7;

prehistoric otter hunting on, xx. *See also* Ainu (Kuril Islanders)
Holmes, Cynthia, 103–4, 122–23
Hooper, C. L., 87, 97, 154n27, 154n44
Hooten, Melvin, 125
hostage taking, 27, 34, 59
Hudson's Bay Company, 50, 59, 62–63, 71–72
Humalgueno people (Baja California), xxi

igax (Aleut canoe), 146n20
Igler, David, xv, xix, 2, 62, 142n18
indigenous people: hostage taking, 27, 34, 59; hunting methods of, 5, 11–12, 28, 73–74, 82, 86, 154n44; impacts of colonization on, 3; impacts of fur trade on, 3, 8, 13, 27, 60–61, 62, 84, 87, 96; participation in fur trade, 27, 35, 44–45, 72, 82, 84, 86–88; prehistoric hunting by, xx–xxiii; romanticizing otters, xix; sea otters in mythologies of, xxii–xxiii; tribute (*yasak*), 2, 9, 12; uses for otter fur, xxii; violent treatment of, 1–2, 33–34, 51–52, 59–61, 144n39. *See also* Native Americans; *specific groups*
Ingraham, Joseph, 61
Itelmen (Kamchatka Natives), 27, 28
Iturup Island (Kuril Islands), 1–2, 14–15, 17, 77, 110
Ivashchenko, Yulia, 110

Japan: in the China trade, 4, 145n53; colonizing the Kuril Islands, 3–4, 14–15, 17; in the fur trade, 6–9, 17, 95; geopolitical tensions with Russia, 3, 8–9, 15–16, 88–89; hunting restrictions of, 98; otter aquaria exhibits in, 117; and pelagic hunting, 96; prehistoric otter hunting in, xx; trade networks with Russia, 13; trade with Ainu, 6–8; treatment of Ainu, 3, 144n39
Jefferson, Thomas, 66

INDEX 183

Jewitt, John, 60
John, Samson, 82
Johnson, Andrew, 83
Jones, Robert "Sea Otter," 105–6, 108
Jones, Ryan Tucker, xxii, 23–24, 28, 39–40, 41, 42, 45, 46
Jordan, David Starr, 95
Juzo, Kondo, 14

Kamchadal Natives (Itelmen), 27, 28
Kamchatka Peninsula: Japanese exploration of, 7–8; otter hunting in, 17; Russian colonization of, 16; Russian exploration of, 10–12, 24
Kamehameha (Hawaiian king), 62
Kayak Island (AK), 25
Keiner, Christine, 79
kelps, xviii, xxi, 109, 114, 126
Kendrick, John, 54
Kenyon, Karl, 83, 100, 108–9, 115, 153n15–16
Khvostov, Nikolai, 15
Kiakhta (Kyakhta) trade, 9–10, 153n8. *See also* China trade
killer whales, xv–xvi, xix, 101, 110, 118–19, 158n52
Kimberly, Martin Morse, 77–78, 86, 91
Kinney, D. J., 107
Kodiak Island: Alaska Commercial Company in, 84; end of otter hunting in, 88; first European settlement on, 14; otter hunting in, 40–41, 84, 86; otter population levels in, 86; Russian colonization of, 33–34; Spanish expeditions to, 38. *See also* Alaska
Kornev, S. I., 17
Korneva, S. M., 17
Krasheninnikov, Stepan, 11–12, 13
Krusenstern expedition, 43
Kunashir Island (Kuril Islands), 1–2, 7
Kuril Islands: American hunters in, 77, 82, 88–90, 91; British hunters in, 88–90; conservation efforts in, 109–10, 145n55; factors in otter survival in, 89–90; Japanese-Ainu trade networks in, 4, 6–9, 16; Japanese colonization of, 3, 9, 16; Japanese-Russian geopolitical conflicts in, 3–4, 8–9, 14–16; otter population levels in, 3–4, 17–19, 155n72; Russian colonization attempts in, 3, 13–14; Russian exploration of, 10–12; Russian hunting in, 1–2, 13, 16, 144n30. *See also* Ainu (Kuril Islanders)
Kushner, Howard, 68
Kutune Shirka (Ainu oral epic), xxii
Kyakhta (Kiakhta) trade, 9–10, 153n8. *See also* China trade

Lady Washington (ship), 54
Langsdorff, Georg Heinrich von, 43
Laxman, Adam, 8
Lebedev-Lastochkin, Pavel Sergeevich, 33–34
Lech, Veronica, xxi
Lee, Molly, 85
Lensen, George Alexander, 144n30
Lensink, Cal, 24, 88, 147n64
Light, Allen "Black Steward," 73–74
Livingston, Kirsten, 159n3
Lloyd, James, 151n83
London fur market, 81, 83, 92

Malaspina, Alejandro, 43
Maquinna (Nootka chief), 51–52, 57, 60
Marine Mammal Protection Act (1972), 112–13
The Marine Mammals of the North-western Coast of North America (Scammon), 81–82
maritime fur trade. *See* fur trade
Marquesas Islands, 61
Martínez, Esteban José, 38, 43, 54–56
Maschner, Herbert D., xxi

McIntyre, Charlie, 82
Mears, John, 53, 54, 55–56, 149n20
Medny Island "Copper Island" (Commander Islands), 90, 98
megafaunal collapse hypothesis, 119
mercury (quicksilver), 35
Merriam, C. Hart, 93
Mexico: extirpation of sea otters in, 83; fur trade in, 34–37, 81; ships in California otter trade, 136, 138–39. *See also* Spain
Middleton, Henry, 151n83
Miller, Gwenn, 34
Milo (sea otter), 103–4, 123
Ming dynasty, 4
mission Indians (Native Americans), 32, 35, 63, 74
Moñino y Redondo, José, 55
Monroe, James, 151n83
Monterey Bay (CA), 32, 78–79, 99–100, 111–12
Monterey Bay Aquarium, 115–17, 121
I Moscoviti nella California (Torrubia), 21–22
Muir, John, 92, 125
Murie, Olaus, 98
Muro, Antonio de San José, 36
Murray, Joseph, 95
The Muscovites in California (Torrubia), 21–22

Nance, Susan, xvi
National Zoological Park (Washington DC), 114
Native Americans: and American traders, 57–61; in the fur trade, 35, 44–45, 57–58, 72, 84, 86–88; hunting methods of, 28, 73–74, 82, 86; impact of hunting regulations on, 113–14; precontact violence among, 61; and the Second Kamchatka Expedition, 25; violence experienced by, 33–34,

51–52, 59–61. *See also* Aleuts (indigenous people); indigenous people
Near Islands (Aleutian Islands), 26, 27, 40
Nerchinsk, Treaty of, 9
Nichol, Linda M., 90
Nickerson, Roy, 119–20
Nidever, George, 73–74, 90–91
Nootka Sound (Vancouver Island): American traders at, 54; attack on the *Boston* at, 60; English settlement plans for, 38; English trade at, 50–52; Spanish-English conflict at, 36, 53, 54–57; Spanish exploration of, 32, 36
North American Commercial Company, 96
North Pacific Fur Seal Act (1966), 112
North Pacific Fur Seal Convention (1911), 78, 93–98
North West Company, 62, 66–67, 71
Northwest Passage, 50, 51, 57
Noticia de la California (Venegas), 21, 30
Nyac (sea otter), 103–4, 123

O'Cain, Joseph, 63
Ogden, Adele, 37, 41, 42, 74, 127–39, 147n53
oil spills, 103, 117–18
opium trade, 79–81
orcas, xv–xvi, xix, 101, 110, 118–19, 158n52
Oregon Territory, xxii, 47, 71–73, 82–83, 153n15
Organ, John F., 121
Otters Holding Hands (YouTube video), 103–4, 122–23
Owings, Margaret, 112

Pacific Fur Company, 66
Packard, David, 115
Pallas, Peter Simon, 10
Pardo, Roberto, 74, 152n100
Paul (tsar), 14

pelagic hunting, 85–86, 94–96, 154n31
Pérez, Juan, 32
Peter the Great (tsar), 10–11, 24
Philippine Company, 35, 36
"The Playful Sea Otter" (article), 107
Plummer, Katherine, 143n18
Polynesian people, 50, 62
Portlock, Nathaniel, 52
Pribilof Islands, 77–78, 84, 85–86, 94–96
Prince William Sound, 34, 70, 104, 108, 117–18
Progressive Era, 78–79, 93–98
Pynn, Larry, 122

Queen Charlotte Islands (Haida Gwaii), xx–xxi, 32, 41, 52, 54

Ralls, Katherine, 100
Resolution (ship), 50–51
Rezanov, Nikolai, 15
Riedman, Marianne, 121–22
Roda y Arrieta, Manuel de, 30
Rogers, Eugene, 90–91
Ross colony (CA), 37, 38–39, 45, 64, 68, 147n58
rotation system (conservation method), 18
Russia: 1821 *ukase* of, 67–71; and the Alaska Purchase, 83–84; conservation efforts of, 18–19, 23, 45–46, 84, 90, 110, 145n55; expansionism of, 13–14, 21–22, 24–26, 32, 33–34, 37–38; geopolitical conflicts with Japan, 3, 8–9, 15–16, 88–89; geopolitical conflicts with Spain, 21–22, 37–39; impacts of colonialism of, 46; international responses to expansionism of, 29–30; otter captivity experiments, 98, 114–15; relations with U.S., 49, 67–71, 83–84, 151n83; Second Kamchatka Expedition, 24–26; trade agreements with Britain, 71–72; trade networks with Japan, 7–8, 13; in the whaling industry, 110–11, 157n25
Russian American Company: and the Alaska Commercial Company, 84; conservation practices of, 18–19, 45, 84; cooperation with American traders, 49, 63–65; expanding eastward, 40–41, 43, 44; hunting in Spanish territory, 38–39; hunting on Urup Island, 17–18; and the *ukase*, 67–69; and the Urup colony, 13–14; use of Native hunters, 39, 44–45, 49, 63–64
Russian fur traders: and the California fur trade, 38–39, 63–65, 73, 133–37, 139; in the China trade, 6, 9–10, 143n27, 153n8; demographics of, 11; hunting in the Kurils, 1–2, 10–13, 17–19, 145n55; impact on otter population levels, 39–45; in the maritime fur trade, 2–4, 26–29, 143n27, 153n8; and pelagic hunting, 96; Second Kamchatka Expedition, 24–26; ships of, 26; treatment of indigenous people, 1–2, 33–34, 144n39

Sakhalin Island, 15, 16
Sanak Islands, xxi, 90
sandalwood, 59
San Francisco Bay (CA), 32, 38, 75
Santan trade, 6. *See also* China trade
Scammon, Charles, 81–82, 91
Scheffer, Victor, 82
Schlesinger, Jonathan, 24
sea cows, 26, 46
Seal and Salmon Fisheries (Jordan), 95
seals (fur seals): Alaska Commercial Company's hunting of, 84; in the China trade, 59; connections to sea otters, 45, 93–97; conservation measures for, 18, 45, 85–86, 93–97, 101; illegal hunting of, 77–78, 87; pelagic hunting of, 94–96

Sea Otter (ship), 51–52
sea otters: aesthetic qualities of, 12–13, 118, 121; aggressiveness of, 121–22; anthropomorphic descriptions of, xix, 107, 109, 111, 119–20; in aquarium exhibits, 104, 114–17; building public support for, 106–7, 109, 111, 112, 118; in captivity, 98, 114–17; charismatic characteristics of, 106–7, 115, 121; diet of, xviii–xix; economic value of, xv, 28; in the *Exxon Valdez* oil spill, 103, 117–18; factors in recovery rates, xix; fears of extinction, 92–93; as a flagship species, 121; habitat of, xiv, xviii, xix–xx, 89–90; images of, *xx, 29, 76, 85, 102, 106, 116*; impacts of atomic testing on, 108–9; impacts of World War II on, 109–11; impacts on nearshore ecosystems, 109; importance of, xiii–xvii; in indigenous mythologies, xxii–xxiii; links to beavers, 50, 68, 126; links to fur seals, 78, 93–97; mortality rates factors, 100–101; natural predators of, xv–xvi, xix, 100–101, 118–19, 158n52; Pacific World links through, xiv, 22; political value of, xv, 48–49; population levels in Alaska, 23, 39–40, 98, 113, 118–19, 125–26; population levels in California, 23, 43–44, 50, 73–76, 78–79, 90–92, 99–101, 113; population levels in Oregon, 72, 82–83; population levels in the Aleutians, 23, 39–43, 45–46, 86, 98, 118–19; population levels in the Commander Islands, 90, 98; population levels in the Kurils, 3–4, 17–19, 89–90, 145n55, 155n72; population levels in Washington State, 72, 82–83; prehistory of, xvii–xxiii; previous scholarship on, xiii–xiv; recognized subspecies of, 141n9; recovery of, 78–79, 84–86, 98, 99–101; romantic descriptions of, 97, 107; translocations of, 108–9, 114, 125–26. *See also* conservation; cuteness (sea otters)
Sea Otters (VanBlaricom), 120
Sea Otters: A Natural History and Guide (Nickerson), 119–20
sea urchins, xviii, xxi
Second Kamchatka Expedition, 11–12, 24–26, 29
Serfass, Thomas L., 121
Serra, Junípero, 32
Seward, William, 83
Shakushain's Rebellion, 8
sharks, xix, 101
Sharpe, Howard Granville, 99
Shaw, George, 28, 29
Sheehan, Brian, 122
Shelikhov, Grigorii, 14, 33–34, 38, 40–41
Shields, James, 41
Shimoda, Treaty of, 16
Shumagin Islands, 25
Shumshu Island (Kuril Islands), 11, 144n39
Sierra Club, 108, 156n14
Simpson, George, 71–72
Sitka (AK), 43, 44, 70, 147n63, 150n63. *See also* Alaska
Sloan, N. A., xxiii
Snow, H. J., 17, 89, 91
Sola, Pablo Vicente de, 39
South Sea Company, 52, 55
Spain: colonizing California, 30–33, 43–44, 65; early dominance in Pacific, 2–3; expeditions of, 32–33; geopolitical conflicts with Russia, 21–22, 37–39; Nootka Sound controversy, 36, 38, 54–57; ships in California otter trade, 131–33, 135
Spanberg, Martin, 24
Spear, Nathan, 73
Stanford University, 115
Stedman, Bert, 120–21

Steller, Georg Wilhelm, xix, xx, xxiv, 12–13, 17, 25–26, 29, 145n11
Stephan, John, 6, 13, 14, 144n36
Stevens, Sadie S., 121
St. Paul (ship), 24–25
St. Peter (ship), 24–26
Strange, James, 58
Sturgis, William, xv, 47–48, 58, 60, 69–70, 72–73, 151n74, 151n83
Susie (sea otter), 115
Sutter, John, 39
Switek, Brian, 122

Taft, William Howard, 96
Takahashi, Chikashi, 4
Tang dynasty, 4
Tezuka, Kaoru, 6
Thompson, Frank Wildes, *80*
Tikhmenev, P. A., 17–18, 87–88, 145n55
Tlingit people, 41, 45
Toba Aquarium, 117
Tokunai, Mogami, 9
Tolstykh, Andrean, 40
Torrubia, José, 21–22
tourism, 104, 112, 121
"Trading Vessels on the California Coast, 1786–1848" (Ogden), 127–28
translocations programs, 108–9, 114, 125–26
Treaty of Nerchinsk (1689), 9
Treaty of Shimoda (1855), 16
Treaty of St. Petersburg (1875), 16, 88
tribute (*yasak*), 2, 9, 12, 40
Tsimshian people, 44

ukase (1821), 16, 49–50, 151n74, 151n83
ulyuxtax (Aleut canoe), 146n20
Unalaska Island, 38, 84
United States: and the Alaska Purchase, 83–84; atomic tests of, 104, 107–9; border with Canada, 47–48, 67; in the eastern Pacific fur trade, 43; fur trade regulations of, 87, 113–14; geopolitical conflicts over pelagic hunting, 94–95, 96; geopolitical conflicts with England, 71–73; relations with Russia during fur trade, 43, 49, 67–71, 151n83; ships in California otter trade, 132–39; Sino-American treaty, 80. *See also* American fur traders
upwelling process, xviii
Urup Island (Kuril Islands): following World War II, 110; Japanese colonization of, 15; otter hunting on, 1–2, 3–4, 17–18; otter population levels at, 145n55; Russian colonization attempts, 13–14; Russian conservation efforts at, 145n55; violence on, 1–2
U.S. Revenue Marine Service (Coast Guard), 81–82, 94

VanBlaricom, Glenn, 78, 97, 105, 113, 119, 120
Vancouver, George, 36, 56–57
Vancouver Aquarium, 103–4, 122, 123
Vancouver Island: American traders in, 54; British traders in, 34, 36, 38; end of otter hunting on, 153n16; otter recovery on, 114; Spanish exploration of, 32, 36, 54; trade impacts on indigenous of, 60–61. *See also* Nootka Sound (Vancouver Island)
Vasadre y Vega, Vicente, 35, 37, 147n53
Venegas, Miguel, 21, 30
The Viviparous Quadrupeds of North America (Audubon), 75, 76

Walker, Brett, 3
The Warm Coat (documentary), 109
Washington State: otter extirpation in, 82–83, 153n15; otter population levels in, 72; prehistoric otter hunting in, xxii; Spanish expeditions to, 32–33
Waxell, Sven, 25

Weyler, Rex, 156n14
whales, xv–xvi, xix, 101, 110–11, 118–19, 158n52
whaling industry, 110–11, 119, 157n25
Williams, Perry, 125
Williams, Peter, 113–14
Winship, Abel, 65
Woodland Park Zoo (Seattle), 114–15
World War II, 105–6

Yakutat Bay (AK), 41, 87, 147n63
Yarnyshnaya Bay (Russia), 98
Young, John, 62
Yount, George, 73

Zilberstein, Anya, 60
Zimmerman, Hannah, 159n3
zoos, 104, 114–17
Zvezdochetov, Vasilii, 14

IN THE STUDIES IN PACIFIC WORLDS SERIES:

California and Hawai'i Bound:
U.S. Settler Colonialism and the
Pacific West, 1848–1959
Henry Knight Lozano

How to Reach Japan by Subway:
America's Fascination with
Japanese Culture, 1945–1965
Meghan Warner Mettler

Sea Otters: A History
Richard Ravalli

Hawaiian by Birth: Missionary
Children, Bicultural Identity, and
U.S. Colonialism in the Pacific
Joy Schulz

To order or obtain more information on these or other
University of Nebraska Press titles, visit nebraskapress.unl.edu.

www.ingramcontent.com/pod-product-compliance
Lightning Source LLC
Chambersburg PA
CBHW022100160426
43198CB00008B/294